Is it not Odd the Creator never Shows?

Joseph Compton

CreateSpace, Charleston, SC

Printed in the United States of America
by CreateSpace, an Amazon.com Company

Library of Congress Control Number
2014911151
CreateSpace Independent Publishing Platform
North Charleston, SC

ISBN-13: 978-1500199869
ISBN-10: 1500199869

Contents

Figures

Recognition and Acknowledgment

It is a contradiction in terms and ideas to call anything a Revelation that comes to us at second hand, either verbally or in writing. Revelation is necessarily limited to the first communication. After this, it is only an account of something which that person says was a Revelation made to him; and though he may find himself obliged to believe it, it cannot be incumbent on me to believe it in the same manner, for it was not a Revelation made to me, and I have only his word that it was made to him.

Had the news of salvation by Jesus Christ been inscribed on the face of the Sun and the Moon, in characters that all nations would have understood, the whole Earth had known it in twenty-four hours, and all nations would have believed it; whereas, though it is now almost two thousand years since, as they tell us, Christ came upon Earth, not a twentieth part of the people of the Earth know anything of it, and among those who do, the wiser part do not believe it.

------------ Thomas Paine

Preface

Several religious faiths believe the Creator communicated to humanity through the narrow filter of a chosen man. In the Jewish faith, for example, it is said the Creator communicated Revelations through Moses and those Revelations are now contained in the Torah. In the Christian faith, it is said the Creator communicated Revelations through Jesus and those Revelations may be found now in the New Testament. In the Islam faith, it is said the Creator communicated Revelations through Muhammad and those Revelations are preserved now in the Qur'an. In the Mormon faith, it is said the Creator communicated Revelations through Joseph Smith and those Revelations were written on buried golden plates.

The Reader may have noted above the use of the term Creator rather than corresponding terms that are often used in various religious contexts (e.g., Father, God, Almighty, Lord, Deity, Divinity, Master, Omnipotent Power, and Prime Mover). This has been purposefully done to free the Reader of emotional baggage that such terms are bound to elicit and to allow the Reader to fairly consider unfamiliar concepts with an open and receptive mind.

The Reader may also notice the use herein of phrases such as "it is said that", "is said to have occurred", and "the Creator is reported to have" to indicate that the subject matter being related is taught by a religion and is generally accepted on faith as it may be of a nature that cannot be proven or shown to be fact. If used in every instance of this sort however, the text becomes tedious and would probably inflict more pain on the Reader than is necessary or desirable. Therefore, and because the Reader is wise enough to mentally insert them when needed, such phrases of origination are sometimes omitted.

On the ceiling of the Sistine Chapel and as shown in Figure 14, the Creator breathes life into Adam by reaching out to touch his finger. Although Michelangelo shows the Creator to have the human form of a distinguished elderly man with a long, flowing beard, there is no way of knowing that to be the case nor what other form the Creator may have or even if the Creator has a tangible form. That was Michelangelo's world. With what is now known of the Universe and its billions of galaxies, it is just as likely the Creator is simply an electromagnetic energy field in which case one would probably refer to the Creator with the impersonal pronoun "it".

When referring to the Creator with a personal pronoun it is thus difficult to know which to select. The only readily available candidates are "he" and "she" which are generally used to denote human subjects rather than a divine one of uncertain sexuality. Choosing "he" would reinforce conventional assumptions that the Creator is manlike in form, e.g., as shown by Michelangelo, even though there are no facts to support these assumptions. This choice would make it difficult for the Reader to impartially consider thoughts and concepts different from those commonly held. Choosing "she" avoids conventional phrasing that automatically triggers assumptions and thus the Reader can more readily consider new and different concepts. And consider, dear Reader, is not the Creator the mother of us all?

The word Revelations has also been introduced above. As used herein, Revelations refer to the Creator's disclosures of her nature, will and purpose as they relate to humanity and as disclosed through human intermediaries such as Moses, Jesus, Muhammad, and Joseph Smith.

Introduction

One can live a full and rewarding life on this Earth without ever asking questions such as "where did all of this come from?", "what is out there in the Universe?", "did we really evolve from other life forms?", or "what is the purpose of it all?". Many humans feel no need to look into the structure of the Universe, our position within that structure, or the origin or evolving nature of that structure. They may not have the time nor the opportunity to do so or are simply content to take life as it comes to them in everyday affairs.

It may even be the case that the happiest and most content among us are those who simply live life day by day and participate fully in its endeavors, struggles, disappointments and rewards without ever delving into questions concerning the beginning, the development and the purpose, if any, of our existence on our little planet and within the greater Universe.

For others, however, it is important and rewarding to study and understand as much as possible of the past and present of our species and of the history, structure, and conjectured future of the Universe. That is, to examine and analyze all that we find about us, all that went before, and all that may follow. These efforts often lead into questions of whether or not there is a Creator responsible for the vast Universe that we observe about us.

Some people are certain the answer to this question is in the positive. They often go beyond this to insist the Creator is as revealed in sacred texts of a particular religion such as Judaism, Christianity, Islam or Mormonism. Not content with this, yet others examine religious writings in exhaustive detail, propose lengthy arguments pro and con about the Creator and her objectives, and sometimes even purposely condition their arguments to defeat and deny opposing viewpoints.

But it is easy to become so involved with the doctrines, beliefs and arguments put forward by proponents of various religions that one does not look beyond them to ask simpler and more fundamental questions. None of these can be more important than the question of why the Creator has never shown herself to humanity. Several possible answers to this question are explored in chapter 1. At the end of this chapter it is noted that even though the Creator never shows, a great many humans remain convinced that she has communicated Revelations to humanity in the past and will do so again at some time in the future. This leads to strange twists in logic which are explored in the remaining chapters of this book.

To begin this exploration, chapter 2 examines the question of why the Creator would always communicate Revelations to humanity through the narrow filter of one man? Chapter 6 then asks why the Creator would always use the narrowest communication channel when disclosing her Revelations. Chapters 8 and 9 respectively investigate why she never addresses a group and never addresses just one woman. In Chapter 11, it is noted that the Creator has never provided a tangible object as proof of her visit to humanity and finally, in Chapter 12, it is noted she never reveals something unknown to humanity at the time of her visit.

Then Chapter 14 questions why the Creator would only visit once in the 5000 years of recorded history. Chapter 15 notes that the Creator has never communicated in our modern times of worldwide communications but, just in case she does in the future, Chapter 17 explores the Creator's thinking when planning her next visit to humanity. Then Chapter 18 takes note of a scenario A in which the

Creator communicates her Revelations to one man on Earth and a scenario B in which men of a male-dominated society originate, over many generations, a story that relates the same event. It is observed that, in our time, there is no way to prove which of scenarios A and B occurred. Either may have been the source of our present religious teachings and we have no way to invalidate one or to prove the other.

Then it is noted in Chapter 19 that the foregoing chapters have assumed the Creator's travels across the Universe are conducted at the speed of light. It is concluded that only the Creator knows if this is a fundamental limitation or if she can exceed this speed or even, for example, pass instantly from one location in the Universe to another. If we were to represent the 4.5 billion year history of our solar system as a 24 hour day, Chapter 21 explains that Homo sapiens would have existed during a bit less than the last 4 seconds of that day. It would therefore seem Homo sapiens ranked awfully low on the Creator's "to do" list.

Chapter 22 estimates that 260 intelligent societies may have evolved on planets within 250 light years of Earth but notes they would have existed at different times across the last 10 billion years of the Universe so that it is unlikely any of them existed at the same time as humans. Finally, Chapter 23 addresses stories that the Creator communicated Revelations to a chosen male by noting that no one other than the chosen one was there at the time of the communication and so no one else can know the story to be true or false. Finally, in Chapter 24 the Creator leaves for another Universe.

The explorations of these Chapters cannot be thoroughly considered without some knowledge of the great Universe in which we live and without knowing something of our most recent ancestors. So these important subjects are reviewed in Chapters 3 and 5 which explore the Cosmic Web and the evolution of humanity. And because the discussions noted above can occasionally seem rather ponderous, pedantic, abstruse, and/or academic, Chapters 4, 7, 10, 13, 16, and 20 discuss some of the same questions in ways intended to prevent the Reader from falling asleep between chapters 1 and 24.

Chapter 1

IS IT NOT ODD THE CREATOR HAS NEVER SHOWN
HERSELF TO HUMANITY? She has never made an appearance upon
this Earth, has never spoken to humans, has never walked among us,
has never streaked across the sky, has never crossed over the ocean,
has never descended a mountain, has never walked a meadow or a
prairie, has never strolled one of our streets, has never sat in the shade
of a tree while speaking with a few humans, has never hosted a picnic
or a banquet, has never played with one of our pets, has never posed
for a photograph, has never been seen at a public event, has never
spoken to a group, has never appeared on television, has never come
into our homes and has never sat with us at one of our tables. In short,
the Creator has never been observed on this planet or this solar system.
Accordingly, we have no concept of her form, shape or substance nor
of her plans or desires and, most important, not even proof of her
existence.

In stating the Creator has never shown herself to humanity we
discount the contentions of individuals that the Creator has privately
revealed herself to them in the past. Such claims are extraordinary and
thus must meet an extraordinary level of proof. It is rational to accept
the statement of one person with regard to usual everyday affairs. If a
man stops us on the street and says it is supposed to rain tomorrow we

take that as probable. We have no reason to think this person is making this weather report up. Similarly, we will generally believe someone if they report a tragedy, e.g., an airplane crash or the sinking of a ship, that occurred somewhere in the world. If we have any doubt we generally postpone it until we can get home and watch the evening news to verify the account.

But we will seldom accept one person's announcement of something as momentous and so unlikely as a sighting of the Creator. This is a happening so unexpected that we doubt it until we can verify it with our own senses or with those of several individuals. In other words, we will take someone's word for events that are relatively common and which have occurred before in history. But we justly demand a high degree of verification for an event as singular as the appearance on this Earth of the Creator.

When stating that the Creator has never been observed on this planet we need to qualify the word never. We have no way of knowing if the Creator showed herself prior to the advent of recorded history. That is, prior to five or six thousand years ago. We can only speak of the times since humans began making written records of their activities, observations, and events. If she appeared to some of the dinosaurs we will never know of it.

And we toss out accounts that lack reasonable substantiation. For example, in the Old Testament book of Daniel it states of the Creator that "thrones were set in place and the Ancient of Days took his seat - his clothing was as white as snow - the hair of his head was white like wool - his throne was flaming with fire and its wheels were all ablaze".

Pretty dramatic stuff - clothing white as snow - hair white like wool - throne flaming with fire - blazing wheels - but we have no way of authenticating these observations. No reliable witnesses to this vision have ever been discovered and facts about Daniel are sketchy at best. It is just stated - take it or leave it. We choose to leave it. That the description depicts the Creator as having an appearance similar to that of an old man does not inspire much confidence either - it does not have the ring of truth. In a Universe populated by billions of stars would the Creator have physical elements of one critter living on just one remote planet?

Humanity has suffered through some difficult times in the last five or six thousand years - wars, diseases, droughts, and hard financial times. Why would the Creator not have dropped in during these times to lend some support, a friendly word, or at least assurance that things will improve? Or could it be we are making an unwarranted assumption? In particular, we are assuming the Creator is aware of our affairs, is interested in them, and is concerned when they are not going well. But what if that is not the case? Perhaps the Creator is not only uninterested but is not even aware of our existence as she indicates in Figure 1.

Figure 1
"perhaps the Creator is not even aware of our existence"

Perhaps she is even unaware of our planet Earth and our solar system. She may not even know of our Milky Way galaxy. After all, there are at least 200 billion other galaxies spread across billions of light years of space. From the outer edges of our visible Universe, it would be difficult to even pick out our galaxy. It would take considerable time and effort to discern it amid the clutter of millions and millions of other galaxies. Would it not be ironic that the Creator is simply unaware of the daily attempts of millions of humans to ask for her blessings and care?

There may be another reason why the Creator never shows and that is that is she is simply not interested in humans. We assume she is privy to our thoughts, fears, hopes and aspirations and that she may even intervene in Earth's affairs to lend assistance to us. But there is precious little evidence to back up these assumptions. She seems to have ignored us. So it is just as likely that she has more important things on her plate and doesn't feel it necessary or important to concern herself with those little people on that remote and insignificant little planet.

After all, natural disasters such as floods, fires, earthquakes and hurricanes, occur on a regular basis and the Creator is apparently content to let them continue. There is no history of her ever intervening in the past to prevent these tragedies. Her interest is them seems remote at best. If asked about it, she might just admit "I don't have the time or interest - I've more important fish to fry".

On the other hand it is possible the Creator is aware of we humans and our little green planet but has a lot on her plate and just can't get here at this time. Look, the old girl has maybe 200 billion galaxies to look after. If she visited 15 or 16 of them each month it would still take her a billion years to peak in on each galaxy. Now how long to check up on the critters of our galaxy when she does finally arrive?

Let's assume there are 200 billion stars in the Milky Way galaxy and assume that advanced life exists on a planet around one out of each billion stars so that there are 200 planets for her to visit. But the life forms won't all be in existence at the same time span of the galaxy so let's assume 10% are in existence at the time of the Creator's visit. If she devotes two days to our galaxy, she can linger for about 2 or 3 hours at each of these planets before being absent for another billion years. So if asked why she hasn't visited our little planet Earth lately, the Creator may simply reply "sorry, but I just haven't had time to get to you guys - but you're on my list - let's see - probably not more than 100 million more of your years - hang in there".

But perhaps we are making another unwarranted assumption. When we state that the Creator never shows, we are assuming there is a Creator. Might we go further in this thought process? Perhaps there is not a Creator. Perhaps she just doesn't exist! At least not a Creator in the sense of a being, a presence, or a physical entity. In our experiences on our little planet we have learned that each object was made by someone or several someones and we are thus conditioned to believe all physical objects have a creator. For example, although we don't know them and can't name them we know there were a group of people that assembled our automobile and another group that built our home or apartment and yet another group that crafted our clothes and our furniture and still another that developed our computer.

So when our attention is directed to an object that belongs to us or to one that belongs to another we can be certain that object was created somewhere by someone or a plurality of someones. We are so conditioned to this principle, that we assume someone must have created the Earth and the Sun and our solar system. We extend this belief to all the galaxies we observe and even to the totality of these galaxies which we call the Universe. In response to any celestial object we observe we conclude that object must have been created by someone. Therefore, when we look out in the eventing sky and see a galaxy of stars we assume a Creator originated those stars and arranged them into that galaxy.

But perhaps this line of thought does not hold when we leave our familiar life on our little planet and step out into the vastness of the Universe with its billions of galaxies. After all, other affairs change when we pass to that scale of the Universe's structure. For example, the movement of objects within galaxies is limited by the speed of light but, as the Universe expands, this expansion can cause the distance between galaxies to increase faster than would be indicated by this limitation. So a scientific limitation that is evident within galaxies is absent on the scale of the Universe.

And if it is logical that someone had to have made all these galaxies, would not we then have to logically proceed further along this line of thought and conclude that someone had to have made the Creator? On the other hand, if the Creator can just exist without the intercession of another then why not the Universe? So reasoning processes that make sense within our Milky Way galaxy may break down out in the Universe. Mental steps that seem to make sense here on Earth may not follow when we peek into the vastness beyond. Indeed, it may be there is no Creator. It is rather a lonely thought but let's face it - we may be all alone.

All we know for certain is we live in a vast Universe and can only see so far out into that Universe so that we have no idea what lies beyond. The portion we can observe we call the Observable Universe to differentiate it from the portion we know nothing about. And as we look about this great Universe we find no one or thing or process that built it. Our attempts to see or hear a Creator are consistently met with failure.

Yet even though there is limited evidence for the existence of a Creator and she has never shown herself to humanity, many humans are nonetheless convinced that she has communicated Revelations to humanity in the past and may do so again at some future point. They feel certain she has an intense interest in us and our wellbeing and is up there somewhere just beyond the clouds. In the following Chapters let us explore some of the breakdowns in logic that arise when we examine the possibilities of some of these contentions.

Chapter 2

IS IT NOT STRANGE THE CREATOR, WISHING TO IMPART REVELATIONS TO HUMANITY, always communicates them through the narrow filter of just one man? Never through one woman, never through a group of men, never through an assembly of women and men, but always through just one man, one man deliberately and purposely isolated from all other humans on Earth. No one else is invited to participate or is even informed of this secret arrangement. Is it not curious she would address humanity through just one man when her teachings are intended for all of humanity?

For all but this one man, the Creator never shows herself. Why would she not have addressed us all directly when it would have been so easy for her to do so? What a momentous moment that would have been? Everyone on Earth receiving the Revelations directly from the Creator. Instead, she always excludes us from what could be the most memorable moment in our history.

Imagine, for example, the Creator arriving for a one-on-one meeting at the end of an epic journey of billions of light years through the cold, dark, vast reaches of space. A journey threaded through enormous black voids between clusters of huge galaxies that are each formed with billions of stars and their countless circling planets. At long last the single galaxy of her mission appears ahead. Even now, however, the Creator's efforts are far from over. She must still travel thousands of light years along a selected spiral arm of the galaxy until she discerns ahead one particular star among thousands and thousands of surrounding ones.

With her long journey now drawing to an end, she voyages inward past the cold outer planets of the Sun until she reaches the planet Earth and then descends to a region near the end of the Mediterranean Sea. Finally the time arrives when she can convey her Revelations to humanity. And at that singular moment in history at the end of her memorable odyssey, she passes her Revelations in private to one chosen man and then vanishes, never to be seen nor heard from again.

Figure 2
"at long last, the single galaxy of her mission appears ahead"

Is this not more than simply odd? Is it not incredible the Creator would complete a spectacular journey of billions of light years across the immenseness of her Universe and then confide her message to humanity through just one man and none other? Yet Judaism states Revelations were passed through Moses, Christianity says they were passed through Jesus, Islam relates they were passed through Muhammad, and Mormonism insists that Joseph Smith was the chosen one to receive the message.

In each case, we are to believe the Creator would complete this grand journey and then ignore millions upon millions of humans on Earth? Not just ignore them but purposely exclude them from her private audience with one man. No one else was invited. Care was taken so that no other men could listen in. No women were even aware of the visit. No elders of society were privy to the conversation. None from the selected man's tribe or community were informed. Only one man of all of Earth's humans was selected to receive the Creator's message. For this was a private affair that excluded all but one of the world's humans yet those same humans were then expected by the Creator to mold their actions and futures in accordance with her Revelations.

The choice of a private meeting would seem to not be by chance as at least four religions tell the same basic story. In each, the Creator excludes all other humans and causes her Revelations to be passed through just one of all people then living on Earth. The Revelations are intended for all or for at least all of a selected people or tribe. But there is not one recorded instance in history in which the Creator communicates directly through more than one man. Not one instance when she passes her Revelations through an assembled crowd or even a small group of three or four.

Instead, she arranges that her Revelations are whispered through the chosen man in isolation from all others. The Creator or her agent discloses her message to humans through the selected man and then lingers no more but simply vanishes from history. Back into the vastness of the black voids between galaxies from where she may journey on to destinations unknown but never repeats her visit to Earth. Because it is reported to have happened in several different religions, this strange action must satisfy some reason or design - but why and for whom?

The Observable Universe contains hundreds of billions of galaxies which are each formed by arrangements of billions of stars. It is certain that a substantial percentage of these stars have circling planets so that the planets of the Universe are almost beyond counting. It challenges rationality to think that intelligent life is restricted to just a single one of these innumerable planets. Surely there are other intelligent beings living on at least a few of the other planets. Would the Creator visit these forms just once and, on that single visit, ignore all but one individual?

Figure 3
"whispered through the chosen man in isolation from all others"

Imagine a story that tells of the Creator visiting a planet X and confiding her Revelations only to Y who is one of millions of individuals living on planet X. Who would believe this story? It seems most unlikely yet we are informed that just one man in the recorded history of our planet received each of the Creator's Revelations. The single most extraordinary events of all time on our little planet home and each is purposely hidden from all but the selected one.

That the Creator never conveys to a group of humans is

especially curious because, having created hundreds of millions of huge galaxies that march out to the Universe's distant outer edges, having created the scientific rules that order these galaxies, and having created complex life on at least one planet of these galaxies, the Creator is necessarily all-knowing and all-wise in concerns of the Universe and, on an infinitely smaller scale, of humanity. Because she created every fiber, every facet, every natural law, and every element of this Universe, there cannot be any thought, concept, nor process of which she is unaware.

Therefore, she would have known that ancient humans were almost completely ignorant of the great Universe of which they were a tiny part. The peoples around the eastern end of the Mediterranean only knew of that small region. They thought they lived on a fixed and limited plain and that everything else, the sun, the moon and the stars, was arranged to look down on that plain. They observed the Sun as it set in the West and watched as it appeared the next morning in the East and they devised various theories about how and where it progressed during the night. It seemed apparent that this was the total extent of the Creator's world and that she was always nearby and could visit by merely dropping down through the clouds.

How were they to know the Earth and the Sun were an infinitesimal part of an enormous galaxy formed by billions of stars and that galaxy was only one of billions that marched outward at least 13.7 billion light years? How were they to know the Creator's visit would have to include travels across at least a portion of that vast Universe? How where they to know she didn't hang out in a permanent home cozily situated just above the clouds of planet Earth?

But the Creator knew and she would also have known of the communication errors that would inevitably follow when her Revelations were directed to peoples of Earth through the narrow channel of a single man. She would know that these Revelations would be distorted as they were subsequently passed by word of mouth through a long chain of human communications spread over vast regions of the Earth. By the time these Revelations reached the cold fiords of the North, the deserts of the Australian continent, the high plateaus of Tibet, and the remotest jungles of Africa, they would be warped almost beyond recognition. Is it not curious the Creator would choose such an error-prone process when she had so many superior alternatives?

The Creator would surely be aware that Revelations imparted only through a selected man would inevitably give rise to skepticism in other humans as to the validity of the Revelations. Humans would naturally question the reliability of one who said the Creator had disclosed directly to he and he alone. Would you believe a man who came up to you on the street and told you he had received Revelations from the Creator? You know you would view this with considerable doubt. Why would the Creator restrict her communications only through this man when she could just as easily convey them to hundreds, thousands or millions and why would she restrict her communications to one location on Earth when she could just as easily reach humans around the globe? If she wished millions to receive her Revelations, would she whisper them to one?

Is it not then puzzling, dear Reader, the Creator would introduce this skepticism when it would have been so easy to avoid such introduction? Deliberate exclusion of all but a selected one is bound to introduce doubt in other humans. In ordinary affairs few would care but many may doubt when it is an extraordinary event – a communication from the Creator.

When one man, for example, states he alone has witnessed a somewhat novel but not terribly uncommon occurrence, e.g., a ball of lightning, a meteorite shower, or the aurora borealis, others are inclined to take his word for it. The occurrence is reasonably common and acceptance of its validity requires only a modest confidence in the man. When one man, however, states he, and he alone, has received Revelations from the Creator, others may be seriously inclined to doubt it.

This occurrence is so unusual and rare – the rarest of the rare - that it would naturally give rise to doubt. In contrast, if each of a large group of humans testified they simultaneously received these Revelations from the Creator, the skepticism would be significantly reduced. The larger the group, the greater the reduction of the skepticism, perhaps even to the elimination of that skepticism. If the group is so wide as to include everyone on Earth, there would be no doubt at all. Surely the Creator of the Universe would know that doubt decreases as the number of contacted people increases and increases as the number decreases. Surely the Creator would know the doubt would be greatly increased if the contact were through just one man.

Knowing all of this, is it not strange an all-wise Creator purposely risks the credibility of her Revelations? Especially, when they are so rarely conveyed and when it would be so simple to forestall the risk? Replacing the narrow communication channel of a chosen man with the broad channel of many would significantly enhance the credibility and accuracy of her received Revelations. It would seem, therefore, the Creator would reach out to as many people as possible. The Creator of the Universe would find it an easy task to contact hundreds or thousands or millions of humans whether they were concentrated in one locale or spread over the entire globe. Yet not one religion in all of history records an instance in which the Creator conveyed to other than just the chosen one. How can it be?

Consider that at the time of each of the Revelations recounted in Judaism, Christianity, Islam, and Mormonism the population of the world exceeded thirty million souls. The world population approximated this level during Moses' life sometime in the 13th to 16th century BCE. It was in excess of 170 million when Jesus lived in the first half of the first century CE and was greater than 200 million during the last part of Muhammad's life in the 7th century CE. By the time Joseph Smith lived during the first half of the 19th century CE, it had ballooned to one billion.

Is it not then peculiar the Creator would speak through just one when she wished to impart her Revelations to at least thirty million humans spread over great regions of the globe? Even if the Creator wished to pass her Revelations to only a preferred portion of humanity, e.g. a selected tribe, as some religions imply, is it not strange she would limit her disclosures to just a selected one of the tribe? If you wished to address a group, wouldn't you speak to more than just one of that group?

Note that at their current state of technology, humans have positioned satellites about the globe so that they can communicate almost instantly with other humans spread around the Earth. Because she was able to create the Universe, it is apparent that the Creator's powers of communication and technology infinitely exceed those of humanity. Accordingly, there can be no doubt she has had, since even before the evolution of humans, the power to instantly reach every person on our little planet.

Then would she really restrict her Revelations to one man knowing this meant they would not reach others in remote areas for centuries to come? It is instructive to recall that humans of the times of Moses, Jesus and Muhammad would not have been aware that the Earth was round nor that other humans lived on the far side of Earth. Moses, for example, could not have known of the existence and location of the Chinese people. It is not recorded that the Creator provided him with such knowledge and he could not have known of it otherwise.

And until just lately, some remote tribes in the Amazon were so isolated as to be completely unaware of other humans and of the outer world. How, then, could the Creator's Revelations reach all humans except by the vagaries of chance communications over exceedingly long periods of time? Even then, the Revelations would surely be greatly distorted by having passed through so many hands.

The direct way of conveying Revelations to humanity at the time of Moses would have been to pass them directly to all of the thirty million people then living. This would have been a simple matter for the Creator that created the Universe. By whispering them instead through only one male, millions of people were denied knowledge of the message as it would have taken, at best, many years for it to have reached distant peoples. Those far from the Mediterranean area were denied knowledge of the message during their lifetimes. Is it is credible that the Creator is more interested in passing her message through one selected member of humanity than in assuring it reaches all members as quickly as possible? What could possibly be gained by this easily avoided delay?

And is it not odd the Creator did not inform Moses, Jesus or Muhammad that the Earth was a globe and did not inform them of the location of other people on this globe to insure these people would receive her Revelations? They were surely ignorant of these important facts and the Creator did not find it necessary to inform them. But without this knowledge how could it be certain the Creator's message would reach all people? It is as though she simply cast her Revelations into the ocean and hoped they would wash up on the far shores of the world's inhabitants.

Restricting her disclosures to just one male in the lands at the east end of the Mediterranean Sea at a time when he was ignorant of other peoples of the Earth – ignorant even of the size and shape of the Earth – meant they could not reach others of Earth's peoples for decades or even centuries. During this time, these people would be denied knowledge of the Creator's Revelations – could this possibly have been the Creator's plan?

And yet we are told this was indeed the Creator's intent. We can then only wonder about the why and the how. We can only try to ascertain her goals. Surely her methods and objectives are beyond our slight powers to fathom. We can only trust that the Revelations in her messages to Moses, Jesus, Muhammad, and Joseph Smith were intended to insure the safety, well-being, and future of we humans. As we are carried around the Sun on that little carriage we call Earth, have we really any alternatives?

Chapter 3

WITH SEVERAL QUESTIONS NOW POSED TO YOU, DEAR READER, it is only fair to give you some time for reflection. But note that these questions can only be considered and addressed in the context of our celestial neighborhood and our descent from our most recent ancestors. One cannot thoroughly judge the relationship between humanity and the Creator if one is ignorant or unsure of the vast Universe in which our little planet is embedded nor can that relationship be fairly investigated if one is unaware of our human family's development over the last few million years. It is thus prudent to now devote this chapter and the next to reviewing our present knowledge of these affairs.

Our ancestors once thought the Sun moved west over a substantially-flat and fixed Earth and then mysteriously reappeared each morning in the east. They were unaware the Sun was just another one of the stars they could see each night. Most of them also assumed humans had been created in their present form and that their history did not date back more than a few thousand years. They had no way to comprehend that they were on a huge globe and that people also lived on the opposite side of that globe. With such limited and distorted views of reality, it was understandable they could not intelligently study the Universe nor form rational opinions about the evolution of their species but could only place their faith in legends and myths handed down through generations. And so the years rolled by.

But you have no excuses because humanity has steadily studied and observed our Universe over many centuries, has analyzed these observations, has drawn conclusions with scientific rigor, and has now provided the intellectual background and knowledge sufficient to permit one to analyze the facts and draw measured conclusions. With this wealth of information now available, let us, in this Chapter, wander through the byways and highways of the Universe and then make an acquaintance in the following Chapter with the tribe Hominini of the planet Earth.

The cosmic web of the Universe is so vast it overwhelms our limited ability to envision it. To even attempt a grasp of this great structure, it is necessary to gain some appreciation of our basic measuring stick for celestial distances – the light year. This is the distance traveled in one year by a light beam that moves at the speed of light which is approximately 186,000 miles per second. Except where they are curved by the presence of strong gravitational fields, light beams travel in straight lines. However, if a light beam could be guided about the Earth it would circle it 7.5 times in a single second. If it then shot away from the Earth, it would be nearly 6 trillion miles away at the end of the first year. This is approximately sixty five thousand times the distance from the Earth to the Sun which provides some sense for the vast distance of a single light year. Yet not even the nearest star of the Universe is within a light year of the Earth.

That star is proxima centauri. Although it is the nearest star, it is a red dwarf and is too dim to be seen with the eye alone. It is one of a three-star system. Another star of this system is alpha centauri A and, at a distance of 4.3 light years, it is the nearest star that can be seen by humans. To an observer on Earth the brightness of a star is a function of its actual brightness and its distance from the Earth. For example, the brightest star in the sky is Sirius even though it is twice as far from Earth as is alpha centauri A. Other bright stars include Canopus, Arcturus, Vega, and Capella. Of the twenty five brightest stars, most are less than 600 light years away but Rigel, Betelgeuse and Deneb are between 1400 and 1500 light years from Earth.

Our Sun, these nearby stars and thousands upon thousands of neighboring stars are spread far out across the Orion spur of our precious home which is a thin barred spiral galaxy with a thickness on the order of 2000 light years and a diameter of approximately 100,000 light years. Because humans on Earth view it edge on, light from this massive galaxy appears as a luminous arc across the sky and it is thus called the Milky Way.

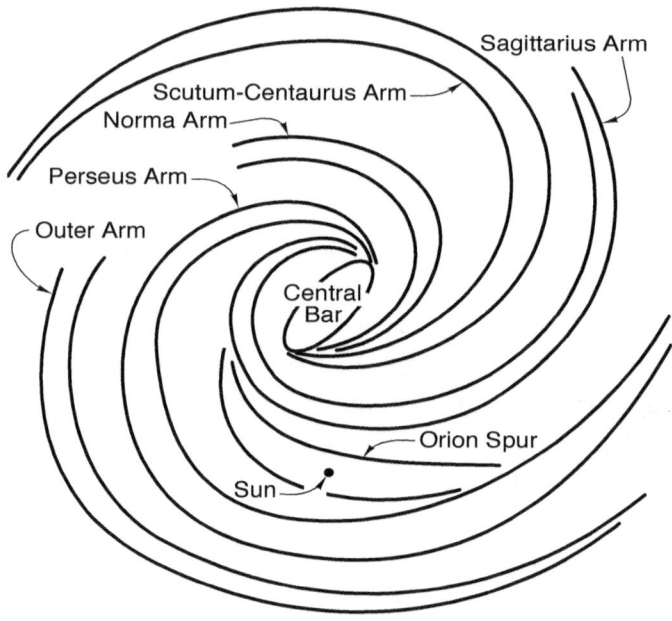

Figure 4
"a thin barred spiral galaxy"

At the center of our galaxy millions of stars are arranged to define a long central bar that probably harbors a massive black hole. From each end of this bar, great arms spiral outward and encircle the bar within. The major arms are the Sagittarius, the Perseus, the Outer, the Scutum-Centaurus, and the Norma. Our little planet is positioned roughly half-way along the Orion spur which is a minor structure that is 3,500 light years wide and arcs for 10,000 light years between the larger Sagittarius and Perseus Arms.

By the most recent estimates, our galaxy has somewhere between 200 and 400 billion stars. Although most of these stars are scattered out along the spiral arms, a great many of them are gathered in huge globular clusters that are sprinkled in a halo about the central bar. The stars in these globular clusters are packed much closer together than those in the spiral arms and they are generally considerably older dating back around 11.5 billion years in contrast to the 4.5 billion year age of our Sun. Their elements are mostly limited to hydrogen and helium with very small amounts of the metals that are common in the younger stars that populate the spiral arms. The Milky Way has approximately 180 of these globular clusters whereas some giant elliptical galaxies are known to have thousands.

Our Milky Way galaxy is one of three large galaxies that together form a galaxy cluster known as the Local Group. The others are the Andromeda and Triangulum galaxies. Also in the Local Group are over forty smaller galaxies which are often referred to as dwarf galaxies. The Local Group is nearly 10 million light years across and contains at least 700 billion stars. Galaxies of the group are held together by their gravitational pulls on one another as they orbit about the center of mass of the group.

Other galaxy clusters lie near the Local Group. These include the Eridanus and Fornax clusters and the much larger Virgo cluster which contains over a thousand galaxies. These galaxy clusters along with two hundred other galaxy groups form the Virgo supercluster which contains at least 200 trillion stars and is approximately 140 million light years wide. About 650 million light years from the Virgo supercluster resides the first supercluster to be discovered. In 1930, the astronomer Harlow Shapely noted "a cloud of galaxies in Centaurus that appears to be one of the most populous yet discovered". Now known as the Shapley supercluster, it measures at least 120 million light years across and contains at least twenty galaxy clusters.

Some of the closest superclusters are the Centaurus and the

Hydra. Then there is the Perseus Pisces which is approximately 300 million light years in length. Further out, more and more superclusters come into view. The Centaurus, the Pavo-Indus and the Coma which is a small, spherical supercluster that is only 20 million light-years in diameter but contains 3,000 galaxies. Then the Sculptor, the Hercules, the Leo and the Pisces-Cetus which is estimated to be a billion light years long and 150 million light years wide - one of the largest structures identified to date in the Universe. And the Bootes, the Corona Borealis, the Capricornus, the Columba, the Sextans, and the Horologium which is also massive with a length of about 550 million light years.

Yet all of this is just the beginning. Galaxy after galaxy, cluster after cluster, supercluster after supercluster, they are scattered like snowflakes across space to form a cosmic web of filaments and walls. There are the Pisces-Cetus and the Perseus-Pegasus filaments and the Sculptor and Centaurus walls. Then there is the Sloan great wall that extends for 1.4 billion light years and there is the 600 million light year stretch of the Coma wall whose entire length is blocked from our view by the gas and dust of our Milky Way galaxy.

Where the filaments and walls of the cosmic web diverge, they define great black voids which are mostly free of galaxies. There is the Local void within our own Local Group. Further out there are the Taurus and Bootes voids. The Taurus is circular in form with a width of 100 million light years whereas the Bootes may be over 300 million light years across. Then there are the Canis Major, Columba and Coma voids. The Corona Borealis, Fornax, and Hercules. The Hydra, the Leo, and the Pegasus. The Perseus-Pisces, the Sagittarius, and the Eridanus. On and on the filaments, walls and voids interweave over and over across billions of light years to lend a filmy appearance to the cosmic web.

The stars within each galaxy are bound together by gravity so that their positions relative to each other within the galaxy are not affected by the expansion of space. The galaxies, therefore, do not grow in size as space expands but they are steadily carried outward and away from each other by this expansion. The galaxies are thus spaced further and further apart with the passage of time. The further a galaxy is from the central galaxy, the faster it is carried outward from that galaxy. This expansion of space is sometimes compared to the expansion of a loaf of raisin bread as it is being baked. The raisins do not expand and neither do the galaxies. But as the distance between the raisins increases so does the distance between the galaxies.

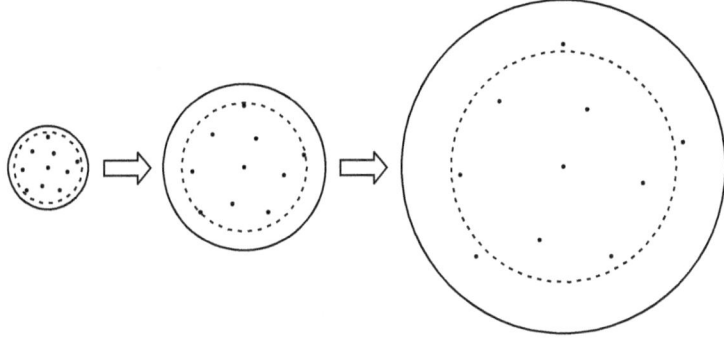

Figure 5
"the central galaxy can no longer observe the outer galaxies"

A broken-line circle in Figure 5 represents the extent of an electromagnetic signal that is radiating outward from the central one of the galaxies at the speed of light. The signal might be, for example, visible light that emanates from the central galaxy. As can be seen in Figure 5, the expansion of space is greater than the speed of light so that the broken-line circle expands at a rate less than that of the solid circle. Thus, the light signal falls further and further behind the expansion of space so that some of the galaxies at the right hand side of the figure are outside the broken-line circle.

This means that these outer galaxies can no longer observe the central galaxy and, correspondingly, the central galaxy can no longer observe the outer galaxies. Light emanating from them will never reach the central galaxy and they have essentially disappeared. If the central galaxy is our Milky Way galaxy, then Figure 5 indicates we can observe only those galaxies that remain within the broken-line circle.

From this discussion, it is clear the expansion of space is not limited to the speed of light. This concept may seem to violate the laws of relativity which constrain structures from moving faster than the speed of light with respect to each other. This restriction, however, applies to cosmic bodies that are bound to each other by gravity. It therefore applies to our solar system and to galaxies, clusters and superclusters but does not apply to space itself which is free to expand at greater rates.

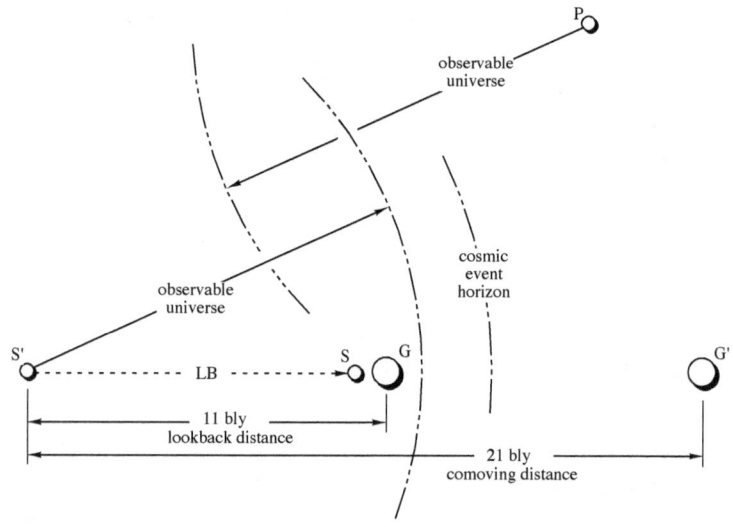

Figure 6
"a light beam LB begins traveling from galaxy G towards a star S"

To further understand the Observable Universe, it is helpful to define the lookback distance and the comoving distance of Figure 6. Imagine that a star S and a galaxy G were 1 billion light years apart in the early times of the Universe. A light beam begins traveling from galaxy G towards the star. As the light beam speeds toward the star it and the galaxy G are being rapidly pulled apart because of the expansion of space. At some moment in the future when the star has reached a final location S', the light beam from galaxy G finally catches up with it.

In an extreme example, the light beam will have been traveling for 11 billion years which is nearly the entire 13.7 billion year life of the Universe. Assume the light beam just caught up with the star yesterday. In this case, the distance between the star's final location S' and the initial location of galaxy G is 11 billion light years. During this vast time, however, the expansion of space has carried galaxy G 10 billion light years in the opposite direction to a final location G' so that the distance between the star's final location S' and galaxy G's final location G' is 21 billion light years.

The lookback distance LB is the distance between the star's final location and galaxy G's initial location. It is the distance that light traveled to the star's final location S' from its inception at galaxy G's initial location. At the moment it arrived, the star could finally "look back" and see galaxy G (accordingly, the arrow LB is directed towards the galaxy G). Prior to that time, the galaxy G was invisible to a viewer orbiting the star S. While the star was moving from its initial position S to its final location S', the expansion of space was moving the galaxy G to its final location G'. In contrast to the lookback distance LB, the comoving distance is the 21 billion light year distance between the star's final location and the final location of galaxy G. It is the distance between the star and galaxy G when the light beam LB finally arrives at the star. The maximum lookback distance defines the 13.7 billion light year radius of the Observable Universe.

Each new year permits light beams from distant cosmic events to have some additional travel time. Accordingly, the radius of the Observable Universe may increase somewhat in the future as more light beams finally arrive at the star after their travels across the Universe. Cosmic events happening now will eventually be visible to a viewer at the star if their initial location permits their light to eventually reach the star. If a cosmic event occurs beyond a final cosmic event horizon, however, the expansion of space will be so great as to prevent the light beams from ever catching up with the departing star. The current distance to this ultimate cosmic event horizon is on the order of 16 billion light years.

The Observable Universe seen from the star at position S' is different from the Observable Universe seen from a planet in a distant galaxy. A distant planet P in Figure 6 will have its own Observable Universe defined by its respective maximum lookback distance. If it is not too distant from the star, the Observable Universes will overlap. Many galaxies in this planet's Observable Universe will never be observed by a viewer near the star position S' and many galaxies in this star's Observable Universe will never be observed from planet P.

Observers on different planets thus see different Observable Universes. It may be apparent that these Observable Universes can march on and outward without limit. The totality of these Observable Universes is sometimes called the Multiverse. We know that space continues to expand and pull the filaments, walls and voids of the cosmic web further and further apart but the extent of the Multiverse is unknown. Observers on Earth will never see those portions of the Multiverse outside of our respective Observable Universe.

In 2004, the Hubble Space Telescope stared for a total of 278 hours at an area of sky that could be covered by one's little fingernail when held at arms length. This area was one thirteen-millionth of the total area of the sky and it was estimated there were 10,000 galaxies in the resultant image. If the imaged area is representative of the galaxy population across the total sky, it would indicate the presence of at least 130 billion galaxies. Light left the oldest galaxies observed only 800 million years after the Universe began. They are thus close to the edge of our Observable Universe but not at that edge.

The Hubble Space Telescope was later outfitted with the Wide Field Camera 3 and the same area of the sky was reexamined in late 2009 for a total of 48 hours. This produced an image of the most distant galaxy ever found. Expansion of the Universe had given this galaxy a departing velocity so high that its radiation was red-shifted into the near-infrared wavelengths. This was the greatest shift of radiation ever measured and it indicated the galaxy was 13.2 billion light years away from Earth at the time the radiation left it. This would have been approximately 500 million years after the birth of the Universe.

It is now apparent that if the Creator and her travel secretary (see Chapter 7) wished to voyage among the galaxies, clusters and superclusters of our local neighborhood, they would have to allocate a considerable amount of time for travel. Assuming they were initially at the edge of Earth's Observable Universe and wished to make visits among some of the local superclusters such as Bootes, Columba, Sculptor and Horologium, they would have to first travel inward across billions of light years.

Once in our neighborhood, it would still require local jaunts of a few hundreds of millions of light years just to stroll back and forth among the local superclusters. If the travel secretary's appointment book did not permit this much travel time between visits, they would have to somehow exceed the speed of light. Since space itself can expand faster than the speed of light, it would seem the Creator could probably manage this. After all, she is the Creator. If not, she needs to learn to kick back and relax because it's going to be a long, long trip indeed.

If a visit to Earth were on their itinerary, they would now voyage inward to the Virgo supercluster. Once there, it would be a simple matter to trip on past the Eridanus and Fornax Clusters and close in on the Local Group. The last few millions of light years would now seem like a stroll in the park to them. Up ahead they would see the beautiful spiral of our Milky Way galaxy. The travel secretary would then probably have to do some considerable searching of local maps to find which spiral led out from the central bar to the Orion spur.

But once they turned onto this local spur they only had to travel about half way along its length of 10,000 light years before seeing the Sun off in the distance. Assuming they avoided all the local speed traps and had no flats, they would now soon speed inward past the stars Deneb and Betelgeuse, then past Canopus and Arcturus, curve around Siriius, take a right at our Sun, and pull up at Earth right on schedule.

Chapter 4

IT SHOULD BE APPARENT THE CREATOR HAS NO
NEED for a manager nor assistants nor a staff nor a board of directors.
This is the Creator, right? She can do anything everything and thus has
no need for the support organizations of a large modern corporation.
For the sake of discussion, however, imagine she could bounce her
thoughts and plans off an appointed communications manager. A
meeting with her manager Chad might have gone something like the
following.

"Yes Chief, you called?"

"Let's get off on the right foot with a bit more respect."

"Whaaa?"

"Chief is a little too familiar."

"Sorry. You prefer "your majesty or your holiness"?"

"Creator will do, Chad."

"OK, Creator it is. Anyway, you mumbled something about
proclamations or manifestations or promulgations?"

"What? – oh - no. Revelations - what I had in mind was
Revelations. Revelations for critters on a planet called Earth –
humanity, human beings, humankind, men, women or whatever they
call themselves - Tess came up with this little level-seven planet for my
next visit - why I really don't know but there you are."

"You want me to write up some Revelations you can convey to

this humanity?"

"No - no, I've already got the Revelations in mind – I'll just use some of those I've used elsewhere in my Universe. Probably the simple standard set I usually use for level seven critters - I generally don't do level seven but things are a bit slow this eon and I could use the practice. No, what I need are plans for the best way to convey them to these beings on Earth".

"Oh, fine, I'm good at that sort of thing. I don't like telling others what to do but once the Revelations are set in stone, so to speak, I have lots of ideas for communicating them. Communications are my thing. That's why they call me the communications manager."

"Good – then, let's get going on them."

"But first, I need to know what the Revelations are intended to do for humanity – what's the intent?"

"Shape them up – make them behave – cut down on riots – give them some goals – keep them quiet and orderly – maintain discipline – keep them home at night and out of the pool halls - that sort of thing."

"Actually Chief, I already did some research on Revelations and learned they are a divine or supernatural disclosure to your subjects relating to their existence, to truth, to knowledge, and to the Universe."

"Really!"

"Yup."

"Can we move on now?"

"Right - OK - well, the first thing I would do is define the territory. I assume Earth is a round globe like most other planets?"

"Duh! Of course it's round! Have you seen a square planet? Chad, have you any experience in this sort of thing?"

"Just kidding – in addition to looking up Revelations I've also done some research on this little planet. It's pretty far out in the Universe but I looked it up in the Universe records. It has beings locally called humans and I found there are significant numbers of these humans in just about any part of the Earth you pick. For example, large numbers are spread across several continents that form a considerable portion of the Earth. Then there are smaller groups of people in other regions that are pretty much all around the Earth – some are isolated in very remote locations that are hard as the dickens to get to. There are a lot of these human thingys. Getting your Revelations out to each of them is going to take some doing – this is not going to be easy – but don't worry, I've worked on other big projects and I've got some thoughts on how to proceed."

"Actually, I was thinking about just communicating them to one man and then splitting."

"Ha, ha, sure. Now here's one way of going about this, Chief."

"I wasn't kidding, Chad and I told you it's Creator."

"You must be kidding, Chief."

"I wasn't but just for laughs, lets hear your thoughts."

"OK, first thing to consider is how to access all of the humans spread about the Earth. Giving them to one human and then relying on word-of-mouth just won't work – it would take forever and your Revelations would be distorted beyond recognition by the time they were even half-way there."

"I kind of like just communicating through one man in the area around the east end of the Mediterranean Sea – the Universe records say the terrain there has a sort of biblical feel to it – makes for a dramatic effect – you know what I mean, Chad?"

"I have no idea what "biblical" means but I can tell you're at least somewhat familiar with the geography on Earth – that's good – you've done some preparation for this ramble. Look, we need a comprehensive approach. For example, we first print up a million or so copies of these Revelation thingys of yours and do a survey to identify every inhabited area on Earth. Just for the survey alone, I'll need a staff of several hundred over at least a couple of years – this is a formidable undertaking – I need workers who are willing to go across glaciers, slog through swamps, climb mountains, brave deserts, and cross oceans. This is like doing a world-wide census."

"What's a census?"

"Forget it – look, this is the only way to get your Revelations to all people with a minimum of errors – it's a lot of effort but it's the only way – we have to hand the Revelations directly to every person around the Earth or it's just not going to work. We can't just tell one man and split – that would be a disaster – the results would bear little resemblance to your original Revelations."

"Egypt, Bethlehem, Jerusalem, Medina, Mecca – they all have a nice ring - you know what I mean?"

"Where are you going with this?"

"I just thought I'd etch my Revelations on some stone tablets and give them to one man in that area."

A silence.

"I had another thought of writing the Revelations on some gold plates and burying them – that also has a dramatic ring to it."

An even longer silence.

"Hello, where did you go Chad?"

"Look, Chief – sorry - Creator, I'm busy - I've got other projects to work on also. If we are just going to horse around, I need to know and I'm out of here."

"OK – it was just a thought – what else have you got?"

"Probably a better way to go is to put your Revelations up where every human on Earth can see them. That way we don't have to tramp with them to every remote corner of the Earth. Just station a few copies in lights in different languages a few miles high in the sky and let the Earth rotate beneath them – all each human needs to do is just look up and there they are – every day – we're guaranteed that your Revelations are visible to all and with no word-of-mouth errors. I like that – it's perfect - we don't need to slog all over the Earth trying to deliver a copy to every last human out in the boondocks."

"What are boondocks?"

"Forget it – just an expression."

"What would hold them up?"

"How would I know? That's your department – if you could come up with gravity, you can surely come up with some way to keep a few copies of the Revelations up in the sky – should be a piece of cake for you!"

"Fine, I not only have to come up with the Revelations but also a way to fix them stationary in the sky – why do I need a communications manager that talks too fast if I have to do all the heavy lifting?"

"I'm good at planning but I'm not a scientist – that's out of my field – however, here's another thought - we could write your Revelations on the face of the Moon as Thomas Paine suggested. Then everyone on Earth need only look up each night to see them."

"You're becoming a pain - I don't like where we're going here – this is getting much too complicated, Chad - you come up with too many plans – let's get back to just telling one man or burying some plates or something like that – one day max and we're out of there – this is far from the only planet in my Universe, you know!"

Figure 7
"you appear in the sky in all your glory"

"If you're through fantasizing now, Chief, try this scheme – you appear in the sky in all your glory – light beams radiating outward through clouds of smoke, rattling thunder, flashes of background lightning, the whole nine yards – then you recite your Revelations in a voice that shakes the forests, ripples the waters, bounces off the cliffs - that'll get their attention - sort of like the Wizard of Oz without the curtains so there's nothing between you and them to corrupt your Revelations."

"The Wizard of what?"

"Never mind, the point is all of these human thingys get the Revelations straight from the horse's mouth."

"Stop with the idioms!"

"OK, try this one. You wanted to bury some plates – well, I'll give you plates! We etch a million platinum plates with your Revelations. You check them for accuracy and then we stuff them in a giant asteroid that burns up in the Earth's atmosphere scattering the plates to fall all over the place."

"Would that work?"

"Well, we need to do some work on the entry physics to make sure the plates come through relatively undamaged. If they're singed or blackened a bit though, that just lends some drama – you said you like drama."

"I like it but it sounds complicated and costly."

"You're limited to a budget?"

"Duh! What do you think!"

"Somehow budget and Creator don't seem to go together."

"Well, think again Chad. That's why I like just dealing through one man and relying on him to pass on the Revelations. You can't get much cheaper than that! Wham, bam, and we're out of there!"

"I notice you keep mentioning a man – do you have a hang up about women? Why not pass your Revelations through one woman?"

"Get real! What guy is going to believe a dame that says she spoke to the Creator?"

"You're right Chief – they don't call you the Creator for nothing!"

Chapter 5

THE TERM HOMININS REFERS TO MEMBERS OF THE
BIPEDAL TRIBE HOMININI which is thought to have originated
between 5 and 8 million years ago. Members of this tribe stand
upright, walk and run primarily on two legs, and are more closely
related to modern humans that to any other group. Significant
relationships within the Hominini during the last four million years are
shown in the graph of Figure 8. In particular, this graph shows
relationships between genera included in this tribe. The oldest genus
shown is the Australopithecus and the youngest is the Homo with the
Paranthropus branching off from the Australopithecus and coming to a
dead end somewhere between 1 and 2 million years ago.

The Australopithecus afarensis species lived on Earth between
3.9 and 3.0 million years ago. Members of this species had an apelike
face formed with a low forehead, a prominent ridge over the eyes, a
flat nose, and a receding chin. Their pelvis and leg bones roughly
resembled those of modern humans and were adapted for a bipedal
gait. Females were substantially smaller than males which had a height
between 3.5 and 5 feet. Australopithecus afarensis hands were similar
to those of humans but their fingers and toes were proportionally
longer and more curved. The cranial capacity was between 375 and
550 cubic centimeters. The body and brain sizes of Australopithecus
africanus were both slightly greater than those of afarensis. In

comparison to a modern chimpanzee, africanus had a similar body size and a slightly larger brain.

Although there is some disagreement, it is generally thought the Paranthropus is a separate genus that evolved from Australopithecus afarensis. Only a single skull survives of Paranthropus aethiopicus. It is topped with a large bony ridge and indicates a brain size of approximately 410 cubic centimeters. Paranthropus robustus and boisei had large heads formed with powerful jaws, flat faces and brain capacities on the order of 540 cubic centimeters.

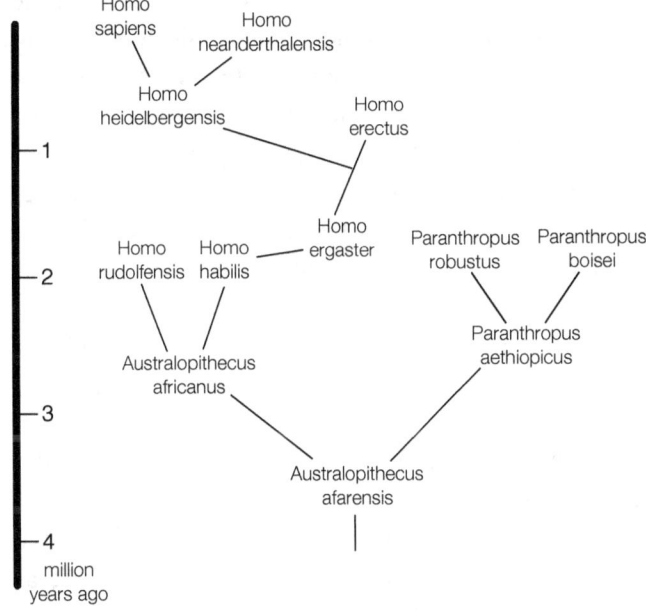

Figure 8
"the bipedal tribe Hominini"

Homo habilis ("handy man") was named because specimens were found along with evidence of tools. Habilis had a height a bit less than five feet, a weight somewhat less than 100 pounds, and a brain size varying between 500 and 800 cubic centimeters. The face projected less than in Australopithecus africanus and the teeth were smaller. It is thought that Homo habilis may have been capable of rudimentary speech. Specimens of Homo habilis exhibit such a wide range of variations that it is believed some should be assigned to the separate species Homo rudolfensis.

It is worth a digression at this point in the description of the genera of FIG. 8 to note the discovery in 2008 of two fossils that led to the formation of a new species Australopicthecus sediba. The partial skeletons of a young male and an adult female were found in a cave in South Africa. These individuals have been dated to about 2 million years ago and they exhibit remarkable transitional characteristics. The shapes of the front of their brain cavities, of their faces, of their teeth, of their pelvises and hips, and of their hands indicate a relationship to a modern species such as Homo erectus. On the other hand, the shape of their feet, the length of their upper arms, and the overall body plan indicate a relationship back to other Australopithecines.

On the basis of these characteristics, it may be that is Australopicthecus sediba and not Homo habilis that forms the link forward to Homo erectus. But because it existed concurrently with Homo habilis, it may simply be that Australopicthecus sediba is the dying remnant of an earlier form and not a direct link leading to modern humans. This illustrates the difficulty of piecing together human evolution but also illustrates the truly remarkable ability of humans to discern their ancestral evolution from the limited evidence of the ancient fossil history.

In any case, the species Homo erectus existed for over a million years with some individuals perhaps living as late as 200 or 300 thousand years ago. Their face had a reduced chin, thick brow ridges over the eyes and a long, low skull with a brain capacity that ranged between 750 and 1225 cubic centimeters. This was the first species to walk out of Africa and it subsequently spread to parts of Asia and Europe. Earlier specimens found in Africa tend to be taller and thinner and are generally assigned to the separate species Homo ergaster which is the first species approximating the same physical proportions and size as modern humans. Homo erectus definitely used stone tools and may have used fire.

Homo heidelbergensis had approximately the same height as modern humans but was somewhat lighter in weight. Their faces were formed with large brow ridges and low, receding foreheads. They had elongated brain cases and a brain capacity that averaged 1200 cubic centimeters. The robustness of their skeleton and teeth was between that of Homo erectus and Homo sapiens. Homo heidelbergensis is sometimes considered to simply be an archaic version of Homo sapiens.

Homo neanderthalensis existed between 230,000 and 30,000 years ago. Neanderthals were shorter but more robust than Homo heidelbergensis. Their skulls had large brow-ridges and a protruding face above a sloping chin. With a brain capacity of about 1450 cubic centimeters, the brain case was longer and lower than in modern humans. Males averaged about 5'6", had thick bones and appeared to be heavily muscled. They made tools and weapons and were the first of the Hominini tribe to bury their dead.

Early forms of Homo sapiens appeared about 195,000 years ago. They are characterized by small eyebrow ridges, flat face, high forehead, prominent chin, tall and slender frame, and a brain capacity of approximately 1350 cubic centimeters which is greater on average than that of Homo heidelbergensis but less than that of some members of Homo neanderthalensis. Within the last 100,000 years, there has been a further reduction in robustness of the teeth, jaw and face. The tool kits of Homo sapiens gained sophistication about 40,000 years ago. Over the next 20,000 years, in addition to tools, they developed beads, artwork, and musical instruments.

Now in our own time. Homo sapiens dominate the Earth and has developed highly complex methods, systems and structures. Even if they are on opposite sides of Earth, members of this species can communicate with each other almost instantaneously through devices such as personal computers and wireless phones. This communication is effected through a vast global network simply referred to as the Internet. In addition, vast amounts of information can be accessed

36   Is it not Odd

through the Internet.

But the Creator could not have used the Internet before modern times. Were she to visit humans before these times, she would have had to used some other communication process. Various religions claim she has effected her communication through one solitary male member of Homo sapiens.

It is interesting to wonder if the Creator ever visited Homo heidelbergensis? Would she have visited earlier members of the Hominini tribe such as Homo Habilis or even Australopithecus africanus? We have no way of knowing because the recorded history of the Hominini tribe is limited to the last 5000 years or so. If she had visited these earlier species it would seem to have been only to observe them because they probably lacked the capacity to communicate with the Creator or to understand her. Of course, we will never know for certain.

How about Homo neanderthalensis? If the Creator communicated to members of our species, there seems no reason why she would not have done the same with the Neanderthals. Surely it would have been as easy for the Creator of the Universe to have grasped their language as it was for her to be able to communicate to Moses, Jesus, Muhammad and Joseph Smith. But again, we have no way of knowing because they left no written records. All we can ever know of them is limited to our study of their bones, tools and caves.

Still, we can imagine the Creator meeting with one of the Neanderthals up on a mountain top. Would the chosen one have been a male? Would the Creator's message have been a list of homilies? Would she have told them to welcome the Homo sapiens? Would the chosen one have had any way to pass on the Creator's message to others? Would the Neanderthals have cared? Would they have had the time or opportunity to care? From what we have been able to discern from the many fractures in their skeletons, their lives were typically short and brutal. They appear to have lived grim, harsh lives and would have had little time for reflection or contemplation. A relationship with the Creator may have been the last thing on their minds but that doesn't mean it didn't happen. We simply have no way of knowing.

Then we come to Homo sapiens. Might the Creator have visited our species prior to the last 5000 years of recorded history? If so, we will never know as no written record could have been generated. During this period there would have been hundreds of generations that were hard pressed to survive accidents, disease, and attacks by animals. We know these ancestors of ours made ritual burials by 40,000 years ago, crafted figurines 34,000 years ago, and invented bows and arrows 22,000 years ago. The first cave paintings date back 30,000 years and by16,000 years ago they had become impressively complex in execution. The knowledge of copper dates back 11,000 years, the cultivation of wheat 10,000 years, and formation of oven-fired pottery 8,000 years. Cattle and horses were domesticated 6000 years before our time. Use of the wheel was evident 5700 years ago and production of bronze was known by 5500 years ago.

We can never know if the Creator visited humanity prior to 5,000 years ago because humans before this date lacked the ability to record events. Since that time, however, writing was developed and we now have access to many written accounts of interfaces between humans and the Creator. For example, some of these accounts tell of Moses receiving Commandments from the Creator on Mount Sinai, others claim Jesus brought messages from the Creator during his short lifetime, yet others say the angel Gabriel passed communications from the Creator through Muhammad throughout the latter part of his life, and still others insist the angel Moroni delivered messages from the Creator through Joseph Smith via buried golden plates.

But in each of these cases, there was no report of other Revelations during the entire 5000 year span. According to each of these religions, the Creator seemed to bring Revelations only once. Is it not curious the Creator would visit Earth just once in recorded history? What of the people that lived before the visit? If the message was important, should they not have had the benefit? And what of the people that lived after? Would it not have been more convincing had messages been delivered again in modern times?

At this point, it may occur to a thoughtful observer that if the Creator did indeed pass Revelations through one isolated man, she or her representative would have necessarily been conversant in the language of that man. Otherwise she could not have spoken to Moses, instructed Jesus, nor sent messages through the angels Gabriel and Moroni. For the Creator of the Universe, however, this may have been an easy problem to solve.

At any rate, is it not suggestive the messages all came before the advent of modern communications? In Moses' time, there was no one to question him when he came down from the mountain. In our time, he would have been surrounded by television cameras, interviewed endlessly by reporters, and displayed and quoted on the evening news. Every aspect of the interface with the Creator would have been explored. If that interface were invented and didn't actually happen, the intense scrutiny would likely have disclosed this. Could that be why such events all date before the time of modern communications?

Try to imagine the millions of years of Figure 8 as they slowly passed by. Would the Creator have peeked in at Earth during the time of Australopithecus afarensis and thought "h'mm, not yet"? Might she have said to herself "I'll check in again later" when she observed Homo habilis more than a million years later? Imagine her as she watched Homo neanderthalensis after yet another million and a half years and concluded "too busy just surviving". And might she have thought of Homo sapiens 30,000 years ago "I'll wait till they can at least grow wheat" and later at 10,000 years ago "maybe when they have the wheel". Was it only when she observed our ancestors had finally mastered writing that she said to herself "now is the time on this little planet Earth to disclose my Revelations - but only through one man – I'm still not fond of the masses. And I'm only visiting once – other generations will have to make do – this is not the only world I have on my watch, you know!"

Chapter 6

IS IT NOT CURIOUS THE CREATOR, IN SEEKING TO CONVEY REVELATIONS to millions of humans on the planet Earth, would pass them through the narrowest-possible communication channel - one that begins with a single one of those humans? The Revelations are thus introduced into the severely restricted entrance of a communication channel with humanity eventually receiving versions of these Revelations from the broad exit end of the channel. As the Revelations progress through the narrow entrance, errors are inserted and it is virtually certain that the Revelations issuing at the exit will differ from those that were introduced. The narrower the entrance, the greater the errors.

In contrast, consider a scenario in which the Creator passes her Revelations directly to each living human. Because of this direct interface, it is assured that no errors would be introduced. It would be certain that the same true message would reach each human on Earth. In this case, the entrance of the communication channel is just as wide as the exit and the widest possible channel would thus be obtained. When the Creator, instead, introduces her Revelations via a single human sources of error are introduced as the Revelations must then pass through a long chain of humans.

There is a parlor game in which a message is passed, one by one, through a line of people as exemplified in Figure 9. When the last person in the line recites the received message as he or she recalls it, it is generally seen to have a variety of introduced errors in the form of deletions, additions, and distortions. The recited message significantly differs from the original and the errors are typically so numerous and so corrupting that they provide considerable fun to the game. The source of the errors is the narrowness of the communication channel. The probability of error is greatly increased because the message is successively passed along this narrow channel to its exit.

Each additional communication is a source of error. Although each person tries to accurately pass on the message they heard, they often fail to do so and instead recite a message that differs somewhat from the one they heard. The game illustrates the typical results when humans try to faithfully pass a received message through even a modest number of communications.

Figure 9
"a message is passed, one by one, through a line of people"

It would seem certain the Creator would be cognizant of this error source. After all, she created humans or had them created via evolution and she would know the accuracy of their communications. She would know that the narrower and longer the c communication channel, the greater the inserted errors. Accordingly, it would seem she would surely guard against a narrow communication channel when imparting her Revelations. She could do this very simply by conveying directly to each human rather than through a selected one. This would remove the communication channel errors completely because each human would receive the Revelations directly from the Creator - "straight from the horse's mouth" as they say. Although the memory of each human may introduce later errors, at least the originally received message would be uncorrupted.

And yet we know from the accounts of several religions that the Creator arranged to have Revelations communicated privately to one man. An early example concerned Moses, another concerned Jesus, yet another Muhammad, and a much later example concerned Joseph Smith. But each passing of the Creator's disclosures through a narrow communication filter offers an opportunity for error. From the viewpoint of the Creator, therefore, there would seem to be no reason for communicating through the narrow channel of a single human.

And yet we are told that the Creator repeatedly chose this communication channel when she passed her Revelations to humanity. Because she created the Universe and is necessarily all-wise, she would have multiple ways of disclosing and would not be restricted to passing Revelations through the narrowest of all channels – that of one man. Because she would know the risk of errors, it would have seemed she would have chosen a wider channel.

Would it not also seem likely that various religions would recount different communication channels? Perhaps, for example, one would tell of the Creator confiding Revelations to a gathered crowd in a meadow, another tell of the Creator successively conveying Revelations to humans scattered about the world's continents, and still another would tell of the Creator imparting Revelations to a gathering of women or to just one woman. That is, would it not seem likely that different religions would recount different communication stories? Yet we are told by many religions that the Creator restricted her Revelations through a single person, the narrowest of all available communication channels, and always chose that person to be a man.

There is no record of a religion that tells of the Creator addressing a crowd of people. But there are several in which a chosen man receives disclosures from the Creator and carries them to the rest of humanity. In all of these religions, a selected male is the sole conduit between the Creator and all other humans. No other human, according to these teachings, has ever been allowed access to the

Creator. The narrowest of all communication filters, that of one man, is all that connects the rest of humanity to the Creator. In these religions, only the chosen one is permitted to listen to the Creator - never a woman and never a group of humans.

Just this one man forms a link between the Creator and humanity. Humans are asked to believe this man who states the Creator has confided her Revelations only through him. If this tale be false, if this link fail, no religion founded on it can prosper or even survive. Without this fundamental pillar, the structure must crumble. All the other stories and accounts that generally accompany religious tales, no matter how intricate or detailed they may be, cannot suffice to support the edifice on their own. This tale, this filter, this link to the Creator can be compared to the trunk of a tree - there may be scores of branches and hundreds of leaves but it is of no matter, without the trunk all the rest fails.

In a quiet moment away from the distractions of the day and upon profound reflection it may dawn on a thoughtful person how vital is this link and this may instill questions about the Revelations. That they could possibly be questioned may arrive like a sudden sunrise out of morning clouds. The truth can be a shock but also revealing. One may suddenly realize how odd it is the Creator would resort to so frail a link to humanity. After starting the intense fusion fires of a million, million stars, after providing gravity to pull these stars into massive galaxies, and after generating electromagnetism to carry the light of these stars to their circling planets, would the Creator then convey her Revelations through one man?

Might there then be some other reason for this process to have been reported by so many religions? Might it be that this is the easiest process to substantiate? If it is said that the Creator, at some time in the distant past, disclosed Revelations through the chosen man, who is to contradict this? What evidence could be brought forth to dispute the claim? By definition, only the chosen was present so no one else on Earth can say the communication did or did not occur. If it is said that the Creator conveyed Revelations to a group of humans, however, each of the group must corroborate this account. This is not a problem if the disclosure indeed occurred but if not, it would be fatal. Could this be the reason so many religions tell of the Creator restricting her teachings to one man? Or has the Creator other reasons not readily apparent to us? Perhaps we cannot see through the clouds of time to realize the truth?

Chapter 7

IT IS CALLED THE MILKY WAY BECAUSE OF ITS APPEARANCE across the sky when viewed on a dark night far away from city lights. More than 100,000 light years across, our Milky Way galaxy, when seen from outer space, resembles a beautiful white flower formed by well over 200 billion stars that are arranged in a broad central bar and multiple arms that spiral outward from the bar. Approximately 26,000 light years from the bar, and positioned on the Orion spur amongst thousands of other stars, is our humble little Sun and its circling planets of which one is our quiet home the Earth.

It is the latest destination of the Creator and her travel secretary Tess as we drop in on them somewhere in the outer fringes of the Observable Universe. Since the Creator is all-powerful and all-wise, it may be assumed she is not limited by the travel restraints of mere mortals and can cruise across the Universe in her Cruiser faster even than a light beam. To appreciate the vast distances, however, let her travels be defined in terms of light years.

"Wow, I thought that last communication of Revelations to those strange level-nine critters with revolving heads would never end – I'm getting on a bit and these things seem to take more out of me every eon – OK Tess, where are we off to now? What's next on your list?"

"Chad scheduled a planet Earth in what its inhabitants call the

Local Group of galaxies. Along with the Fornax and Virgo Clusters, it's all part of the Virgo supercluster – it's at least 800 million light years beyond the Bootes supercluster so you can lean back and relax for awhile. Even in with our Cruiser in ultradrive, it'll take a spell but this is one of the smaller planets so once we arrive I don't envision us being there that long – I'm trying to follow Chad's plans and arrange some mass meetings with maybe a few hundred thousand of what they call people or humans and then we're out of there."

"Humans – are those the creepy, crawly critters with lights on their heads?"

"No, they're in the Fornax Cluster - people of Earth are arranged with two legs at one end and two arms and a head at the other – a somewhat primitive plan but they're coming along nicely – intellectually, they're level seven."

"Oh, right! I forgot about them. Awhile back when things were a bit slow I told Chad to schedule this visit. In Cruiser's library I learned that these people think I never show myself. That I always send a proxy. So they think I don't exist. They think they are such smarty pants."

"Well I think it would have been better in the past if you had shown up in person - and why not?"

"It's easier sending a stand in - this way I don't have to put on the big robe and get ready to appear immortal and all that."

"I've never seen you without the big robe - I thought it was part of you. But that reminds me that I wanted to call your attention to some stains that really should be laundered out."

"Forget it Tess - I'll hit them with a lightning bolt from my finger - that'll do it. But back to my Revelation thingys - they are really wasted on level seven. From now on, we need to keep these affairs to level eight and above. Level seven – oh boy – OK, I'll just meet with one of them and then we'll split."

"With all due respect, Creator, Chad's research indicates that really doesn't work too well. Most of your level seven critters have short memory spans and poor communication skills and when we have checked back in one or two eons, we found their understanding of your Revelations to be seriously flawed when they were disclosed to only one of them.

"Look Tess, at level seven it really doesn't matter if things get a bit garbled. At that stage of development, it's all pretty much the same so let's just restrict the communication to a member of the slower one of the sexes – preferably the males. They usually ask less questions. By the way, what do you think of her by now?"

"Pardon?"

"How does she handle? Pretty neat, huh?"

"Sorry Creator, I'm not following you."

" Hummer's baby - the new Cruiser - what do you think?"

"Oh yes, I didn't know what you meant! No, she's really something. Smooth as silk. I had a little trouble getting used to the new controls and all the new instruments on the observation deck but now she's right as rain. It was always hard to get Old Hummer into ultradrive - she kept slipping back into superdrive and the lurch nearly threw us overboard. And the eternal hum for which we named her - used to drive me a bit batty and I heard you complain more than once. But this beauty just goes and goes and is silent as a ghost. How did you put her together?"

"I didn't - I got her at the Lynx supercluster out near the edge of this Observable Universe. You can hire some really low-cost labor out there and I already had all of the plans and specifications in my head. I was sick to death of that eternal hum of Old Hummer so I designed this baby to be quiet. And I got a guarantee on any problems through the first 10 million years."

"Somehow I never envisioned the Creator as a customer - especially one receiving a guarantee on something - I thought it worked the other way."

"Well guess again - I can drive a pretty good bargain when I want to. I could have built her myself but I'm not quite as spry as in eons past. Long before your time I didn't use a ship at all - just zipped back and forth in my robe - you probably didn't know that - but it got too chilly and I gave that up."

"You mentioned it once or twice before. By the way, what did you do with Old Hummer?"

"Traded her in on this beauty - now she probably has a new paint job and is hauling around a bunch of critters in some remote cluster."

"Well anyway, I absolutely love looking up at all the galaxies through the windows here on the observation deck. And the library is superb - you know how I love to spend hours reading about the Universe. I fixed some sandwiches in the galley and then slipped in there and had a great time looking through everything known about this so-called "Local Group" - the cluster we're headed for. By the way, I see the Bootes and Ursa Major superclusters up ahead. We'll shoot right between them soon so it won't be too much longer now."

"Sounds like we're well ahead of schedule."

"And that agrees with the instrumentation on the observation deck that indicates superdrive on this ship is significantly faster. With Old Hummer it was getting harder and harder to get around. But a lot of that was also because the Observable Universe keeps expanding. Did you ever consider keeping Old Hummer and just shrinking the Universe instead?"

"Odd you should mention that Tess - I actually intended that but when I threw a constant in my space-time equations I rounded it off to a couple of digits. Who knew that would make such a difference? Now the Universe gets bigger each eon."

"So why not just make equation corrections now?"

"I'm not as swift as I used to be back in the day - I'd have to go back through all the calculations and then I might still not get it right - easier to just get faster Cruisers."

"I don't think we should talk about that too much - most critters would be put off to know their Creator can't make some simple space-time corrections."

"Simple, schimple! You try them Tess! Carrying all those digits forward and keeping things straight gives me a headache and a half. It's a pain in the butt if you know what I mean."

"But what happens when space has expanded so much it is basically empty and dead - that doesn't have a very happy sound to it."

"I agree but what do you want? Still, I may do it yet - I'll have to set down and get all my papers back together."

"I'll help if you want - I can put some tables together in the observation deck and brew up a couple of gallons of coffee."

"What's coffee?"

"Something they use on this planet Earth when they have to do a lot of mental work - I was reading about it in the Cruiser's library and I downloaded the ingredients as part of our background work for this visit."

"Good thinking but not right now - my feet hurt and I'm just not in the mood."

"OK - well, in any case, the Virgo supercluster is coming up ahead and I can make out the Local Group. By the way, the galaxy we're looking for there is called the Milky Way."

"Oh, that's a quaint name – level seven all the way."

"They're all your critters, Creator and they do the best they can."

"Good for you Tess – you're always the sympathetic one. Is that it - the barred spiral galaxy up ahead?"

"Yes, and isn't it gorgeous? Even though I've seen thousands of spiral galaxies in all of our travels, I'm still impressed - what a beauty! So now we'll just curve in past its central bar and then we'll zip out the Sagittarius arm to the Orion spur. Wow, it always amazes me how many millions of stars you managed to pack into these babies. I can make out a couple of big ones up ahead - Deneb and Betelgeuse."

"Yes, I was pretty good at it back in the day. OK, remember now – I'm meeting just one of them. Maybe I'll give him one of those sets of stone tablets or the metallic plates – level seven always goes for those in a big way – no use asking them to perform above their level. By the way, let's shift down to the speed of light – you know I don't like that sudden deceleration at the end. I had a sore neck for days after that last visit out in the Sloan wall."

"Are you kidding, Creator? You never seem to get the hang of the distances. If I shift into the speed of light now it would take us thousands of years just to voyage out the Orion spur to this planet Earth. I didn't bring along enough reading material in the library or sandwiches in the galley for that. We'll just shift down from ultradrive to superdrive until we get within about two hundred million miles."

"OK, you da man, Tess. Ha, ha, little humor there. By the way, what are years and miles?"

"According to the Universe records on my computer, a year is the time for the Earth to circle its Sun and a mile is a distance unit used by these humans. OK, we're swinging in now past several local stars – Antares, Canopus, and Arcturus - so the Sun should be coming into view soon – is that it? No, that's Sirius – ah, there it is."

"I see it - I hope you know what you're doing – this arrival part always makes me nervous."

"Don't worry - I'm good at it - I've had lots of practice - now we'll shift down into the speed of light and, let's see - it's supposed to be the third planet out – ah, yes! – that's it – that checks out with the Universe catalog in the library - the pretty little blue planet up ahead. Look, I still don't like your decision. I just confirmed your private meeting alone with one of their males but it's not too late – I can still set up meetings with people all over the Earth. We can pop around it in minutes."

"Skip it Tess, one male it is and then we're out of here. My feet hurt and I'm tired of all these galaxies – there're all pretty much the same. I really got too carried away creating these things back in the day. Anyway, these critters are just level seven and all I can think of right now is a nice, long nap."

"It seems like an awfully long journey to convey to just one male. Would it really hurt you to speak to a female once in awhile? Or how about a short meeting with all the elders? Do you know how long will it take for this one male to pass your Revelations on to all the other humans on Earth?"

"Look, Tess, we've been through all that. The females ask too many questions and get pushy and I don't like groups – they always want to give me the keys to something or other. Anyway, if they were level eight or nine, it would be different but at level seven a few communication errors aren't going to make much of a difference one way or the other."

"I recall there's a really neat astronomical body circling Earth – let me look to see what the humans call it –ah yes, here it is, they call it the Moon – you could burn your Revelations into the face of the Moon and everyone on Earth would be able to read your Revelations forever and there would be absolutely no doubt the Creator had visited. I see on my Universe records here that one of them, Thomas Paine, actually proposed that years ago – pretty smart for level seven I'd say."

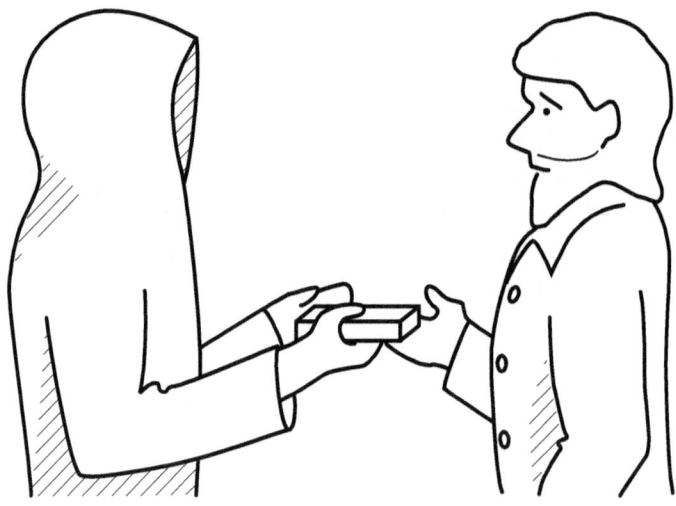

Figure 10
"formed from an element that doesn't exist on this little planet"

"No, I've been all through that with Chad. - did you get that idea from him? Anyways, I did that once out in the Hydra Supercluster a few million years ago – before your time. It takes one heck of a lot of energy and it almost burned off the tip of my finger – I'll not do that again."

"At least leave something so these human critters will know in ages to come that it was the Creator. Why not leave an inscribed vase or a cup or a bowl or something like that formed from an element that doesn't exist on this little planet? When they advance to level eight, they'll realize it could only have come from the Creator – it will prove you made this visit."

"Cheapens the whole thing if I have to prove it, Tess – can't explain it but I've always wanted to be believed just on my own. Anyway, I don't know how many of my elements are on this little planet. That was a long time back when I formed things here in the Virgo supercluster. I'm tired, it has been a long day and I don't feel like looking it up in the Universe records."

"Still I hate to rely on just one male – you know the communication skills at level seven."

Suddenly, Tess saw a blinding flash of light and knew from long experience the discussion was over as the Creator had transported herself to the Earth. Because the only thing to do now was to await her return, Tess began making arrangements for the remainder of the Creator's present schedule. They were due next out in the Horologium supercluster – a good 900 million light years from Earth. She had hardly begun when the Creator reappeared in another blinding flash of light.

"Can't you turn down the transport energy level a bit, Creator? That's really hard on the old retinas."

"Sorry Tess but I can't wait to get out of this galaxy. Cancel everything - I'm through for the – uh, what do they call it around here – oh, yeah – I'm through for the day. Let's blast out of here so I can kick back and relax."

"How did it go? You disclosed your Revelations? You warned them about idols, adultery, coveting and all that stuff?"

"How many times I have to tell you, Tess? It's not just stuff – I worked a long time coming up with this list. Remember how big it went over out in the Bootes supercluster? Anyway, I met a tall, nice-looking male with a beard and he seemed to understand them. Now it's up to him to pass them on to all the other Earthlings."

"Has he done any traveling? Does he know where everyone lives on this Earth?"

"I don't know. I didn't check."

"Does he have any way of traveling to other areas? How will he reach other people? Does he even know the Earth is round?"

"What is this, Tess? Twenty questions? If you stick to the travel arrangements and I stick to passing on my Revelations we'll go far – otherwise, there may be some changes made."

"I'm just trying to make sure your Revelations get out to all of these humans. Seems a shame for us to come this far and not reach everyone."

"It's OK – we'll check back in a million years or so when they reach level eight."

"I just feel for them - stuck with that primitive body plan – they probably need all the help they can get."

"Always the concerned one, aren't you Tess? – hey!, I remember this neighborhood now – there's a nice little spot up ahead in one of the Sculptor superclusters – that's maybe 600 million light years – if we push it in ultradrive we can be there in time for Happy Hour!"

Chapter 8

IS IT NOT CURIOUS THE CREATOR HAS NEVER, EVER COMMUNICATED TO A GROUP OF HUMANS? The long years of recorded history recount several instances in which the Creator communicated through one isolated man but never a single occasion in which she communicated to a group. The Creator is said to have conveyed to humanity through Jesus. It ins also said the Creator spoke through Moses hundreds of years earlier and through Muhammad hundreds of years later. And it is said the Creator informed humanity through Joseph Smith in the nineteenth century. But never has the Creator addressed a group of humans on our Earth.

It cannot be that the technical skills of the Creator are limited as she has created the roaring thermonuclear furnaces of innumerable Suns, organized billions of those Suns into great galaxies, gathered galaxies into vast clusters and superclusters, and arranged them all into an enormous cosmic web that extends across billions of light years of the Observable Universe. Accordingly, it would have been the merest trifle for the Creator to have provided her Revelations to several humans or a large group of humans or to all of humanity.

And the Creator might adopt various tangible forms For example, she might appear simply as another human. Or she might appear in various nonhuman forms such as a blinding burst of light energy. Yet again, the Creator may not be tangibly visible but her disclosures are delivered in a great, booming voice, or are burned into the side of a mountain, or are arranged as text across the sky to thereby shimmer in mid air, or are scanned about the horizon so that there could remain no doubt it was the Creator.

Imagine an instance in which the Creator appears to a man and a woman and speaks to both at the same time and place. Each would then have been able to pass on the revealed Revelations and each would have been able to provide testimony that the other also received the Revelations. Because these testimonies would have corroborated each other, they would have been of great value in assuring a skeptical humanity that the Revelations had indeed been received from the Creator. A man and a woman who can each testify that they together listened to the Creator are far more convincing than one man who insists the Creator spoke only to him.

The testimonies would be even more persuasive if they came from each of a number of men and women. For example, imagine an ancient scene in which an assembly of men and women are scattered across a grassy meadow and are in rapt attention as they listen to their Creator. Afterwards, each of the assembly could separately disclose the Revelations to others so that a great fan of accounts spread outward like ripples on a lake. In response, ancient historians could have provided written accounts based on the direct testimony of numerous of the original assembly. Because of the multiplicity of original sources, there would have been little doubt in succeeding generations that the Creator had indeed communicated to the peoples of Earth through this assembly. Doubt would decrease as the number of original sources increased. However, this scenario is not part of any religion of the world.

Imagine another scenario in which the Creator would utilize some of her scientific rules that govern the electromagnetic spectrum. For example, the aurora borealis and aurora australis in the northern and southern skies are natural light shows that are generated when photons are emitted from the Sun as solar wind particles and they then collide with nitrogen atoms in the upper atmosphere. In a more complex variation of these natural disturbances, the Creator would have found it an easy matter to electromagnetically generate a great light show in the sky that would span the globe and be simultaneously visible to the last Eskimo at the northern extremes of Greenland, the furthest Indian in the dark reaches of the Amazon jungle, the most southern native at the tip of Tierra del Fuego, the most isolated tribesman of the Sahara desert, and the highest Tibetan on the roof of the world. Within the space of each day, every human on Earth could look overhead and personally view Revelations which must surely be of paramount importance since they came from the Creator.

This light show would have been particularly impressive after sunset each evening. Every 24 hours, the Earth would rotate beneath this display so that its message would be presented to all of Earth's inhabitants. All they need do is look up. All doubt as to the origin of these disclosures would be instantly erased and their credibility would be established for all time. There would be no doubt. The light show would be remembered across the ages and recorded so that future generations of humanity would have absolute certainty that the Creator was the source of the disclosures. Should the Creator think it warranted, the Revelations could even be updated and repeated from time to time. This scenario, however, is not a part of any religion of the world.

Visualize yet another instance in which the Creator suspends a mighty monument in the sky to display her Revelations to the ages. It might, for example, comprise a massive block of titanium into which the Revelations are deeply engraved. The block is permanently suspended one hundred thousand feet above the ground so it is always viewable but never violated or altered. The message is brightly illuminated by light which is permanently directed at the block. The suspension is achieved by a local alteration of the Creator's law of gravity and the light is provided by redirection of light from the Sun.

Preferably, movement of the block slightly varies from the Earth's rotation so that, over a few days, the block moves northward and then southward so that it is visible from all portions of the Earth.

Although the alteration and the redirection would seem to violate the laws of the Universe, this is, after all, the work of the Creator, and nothing is beyond her powers. Such alteration of her natural laws would be a trivial exercise yet neither this scenario nor a similar one has ever been recorded in the long history of humanity.

In another scenario, the suspended block can be replaced by a much larger natural body that has been available long before the advent of humans. The Moon is spaced away from the Earth by a balance between gravity urging it inward and centrifugal force urging it outward as it circles the Earth. The Moon is almost perfectly situated to serve as a giant billboard as one face is permanently oriented towards Earth. This is a result of "tidal locking" wherein the Moon and the Earth are slightly stretched towards each other as the Moon orbits the Earth. This effect causes the ocean tides on Earth and also locks a predestined face of the Moon into a permanent orientation towards the Earth.

Full Moon occurs roughly every 29.5 days and at this time the face visible from Earth is brilliantly lit by the Sun. As noted in the recognition and acknowledgment portion of this book, the availability of this natural billboard-in-the-sky was recognized by Thomas Paine whose famous pamphlet "Common Sense" was critical in the initiation of the American Revolution of 1776. Is it not far more likely the Creator would have utilized this natural billboard than secretively speaking through one lone man? Is it probable the Creator would have created doubt by limiting her Revelations to one man rather than displaying them over the face of the Moon for viewing over time immortal? Is it likely she would have encouraged doubt when she could have realized certainty with this natural billboard?

It is estimated that the number of stars in our Milky Way galaxy is far in excess of 200 billion so that, in yet another possible scenario, the Creator might have arranged a few thousand stars, i.e., less than one in each million, so that they spelled out the Creator's disclosures to all of humanity for all eternity. The message could have been duplicated (and written in different languages) to be visible from each of the northern and southern hemispheres and would last essentially forever. It would be visible each evening to each member of humanity and could never be altered by humanity.

The Creator's Revelations would thus bypass the narrow communication channel of a single human and would instead be passed through a direct link between the Creator and each member of humanity. Knowing the frailty of human communications, this would seem a logical solution. What better billboard in the sky could ever be imagined? There would be no doubt as to the source of this message as it could have been generated only by the Creator. It could never be altered and would never be doubted. It is apparent to all, however, that neither this scenario nor a similar one has ever occurred in the history of humanity.

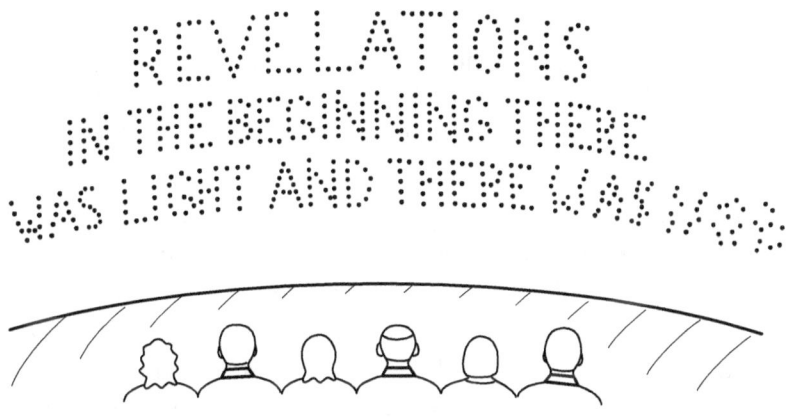

Figure 11
"the Creator might have arranged a few thousand stars"

These are but a few of many scenarios that could have been devised in which the Creator would convey Revelations directly to humanity without passing them through the narrow communication channel of one man. In the first scenario, the Creator's disclosures would have fanned outward from many, if not all, of the original assembly so that doubt as to their credibility would have been significantly less than had they originated through one man. In the last three of these scenarios, the disclosures would have been directly conveyed to each human on Earth so that their source and credibility would have been unquestioned.

Is it not strange the Creator would select a less credible method to pass her Revelations to humanity? Would she risk the skepticism that is inherent when a selected one states he, and only he, received them? Why would she rely on this narrow fallible communication channel? After all, anyone can state without fear of contradiction, that they and they alone received Revelations – by definition, there is not an independent observer to dispute the claim that the Creator spoke to they alone. Of course, neither is there anyone to corroborate the claim. It seems, therefore, a contradiction of the highest order that the all-wise Creator would pass her Revelations through a selected man when she had so many alternative options that would have been more credible. Is it not likely the Creator would convey directly to each human rather than to the selected one?

It may be appreciated, however, that this Revelation scenario may find adherents as it offers several advantages. A first advantage is that it cannot be challenged by other humans since there were no observers of the communication. The veracity of the Revelations can be stated but not validated and must, therefore, be accepted on faith alone. Facts can be challenged whereas faith cannot. Another advantage is that this scenario finds a ready audience among the men of a society since it reinforces their primacy in the society. Yet another advantage is that it establishes, for a race or tribe of humans, a special relationship with the Creator – one not shared by other races or tribes. This scenario can be crafted to lend legitimacy, authority and power to one group of humans relative to others – after all, who can doubt the word of the Creator? These advantages are lost when the she speaks to all of humanity.

And so we are told that Revelations were passed through a selected man and not just once but on at least four dates in the past. The story that she chose just one man need not be one consciously and deceitfully generated by humans. Rather, this story may have just gradually grown over time and through word-of-mouth so that the process was not perceptible to those of any one generation. As the story was repeated over and over, it was gradually enhanced yet each person felt he or she was simply learning the accepted faith of all the generations that had gone before. Details of its origin, if known at all, would fade over time so that each generation would be inclined to accept and believe the Creator had indeed selected a single one of their male ancestors.

This process, imperceptible to any one generation, could have produced the final creation story over many generations. The process would, therefore, roughly correspond to the principles of evolution wherein a variation step is succeeded by a selection step to gradually produce change. In this case, the variation step is provided by believers who add, alter or embellish story details in successive generations and the selection step is provided by followers who choose among these details to fashion a story which most effectively blunts and softens the harsh realities of life with the promise of a caring Creator. And yet vast numbers of humans believe Revelations were received from the Creator through the conduit of a single man - is it likely so many can be in error?

Chapter 9

IS IT NOT STRANGE THE CREATOR NEVER
COMMUNICATES Revelations to a woman? According to several of
the world's religions the Creator has conveyed Revelations through
one man in isolation from all others. From the Creator's viewpoint,
however, there would seem to be no reason for each of these religions
to follow the same story line. First of all, why would she always
restrict her wishes to one person? Given, however, that she wished to
communicate through a single human, it might seem the Creator would
be as likely to choose a woman as a man.

In times past, many societies favored men for positions of
influence and power but would not the Creator operate in accordance
with her own rules? She is not just of this Earth but rules over the vast
Universe where there are probably many advanced societies with
various social rules and customs. In addition, had she conveyed her
Revelations to one woman the prestige bestowed by this action would
likely have overcome any local customs and prejudices.

And the Creator must have her own preferences for
communicating her Revelations. In her vast Universe, therefore, it
seems unlikely she would always prefer one sex over the other when it
came to communicating her Revelations. Actually, it would seem she
would pass her Revelations to everyone on a planet. But if she were to
choose one person to receive them would she really always choose a

male?

Yet Judaism states the Creator communicated only through Moses, Christianity states the Creator communicated only through Jesus, Islam states the Creator communicated only through Muhammad, and Mormonism states the Creator communicated only through Joseph Smith. Thus four different faiths tell us the Creator has channeled her disclosures only through one selected man and never through a selected woman. It would seem, therefore, that the Creator favors men when she selects one human, and one only, as a communication channel.

In earlier times, of course, it was natural to believe humanity was the only advanced society in the Universe. Because even as late as the times of Joseph Smith, most humans were unaware of the incredibly huge Universe that exists outside our solar system. Accordingly, they assumed ours was the only planet that would warrant a visit from the Creator. And because most societies on Earth have been patriarchal societies it was probably natural for them to assume the Creator's attentions would have been channeled through a selected male.

But now we know the observable Universe contains more than 200 billion galaxies which are each formed with billions of stars so that there must be a significant number of advanced societies for the Creator to visit. Each of these societies would have its own unique set of social customs. But would the Creator necessarily observe them when disclosing her Revelations? Is it logical she would rank members of these societies in accordance with the current local social mores or is it more likely she makes her own decisions and ranks all of her children equally? One would assume the Creator marches to her own drum, so to speak. Voyaging down to Earth through the great cosmic web of the Universe, would she then select her own communication process or be influenced by one preferred by local custom?

Imagine the Creator had disclosed Revelations to a woman in one of Earth's regions. Would the women there then be as subservient to the wishes of the men as they presently are? Might they not now use their special relationship to the Creator as a wedge to put forward their own priorities? Once the Creator has spoken through one of your own, there is bound to be increased pride and less reason to be subservient to viewpoints that differ from yours. A special relationship to the Creator would be a powerful social tool that strongly urges deference to your views. If the Creator selected one woman, that would significantly elevate the status of the women of that society.

Figure 12
"the Creator has never conveyed Revelations to a woman"

Then how did it come to pass that the Creator has never conveyed Revelations to a woman of the Earth even though at least one half of Earth's inhabitants are women? Could it have been because the male members of a patriarchal society would prefer to preserve their favored status? For them, it would obviously be of considerable advantage if it were acknowledged that the Creator passed her Revelations through a chosen male of the society.

Since the Creator communicated to a selected male, she is seen to have recognized their authority and given credence to it. It is difficult to go against the Creator's actions. How could a woman now express an opinion about how society should be managed if that conflicted with the opinion of the society's men? Didn't the Creator disclose through one of them and not through one of the women? That should settle that question of priority for a century or two at the least.

Perhaps, then, men would be likely to encourage a story in which the Creator dealt with a chosen male and unlikely to encourage a story in which the Creator dealt with a chosen female. But how can we determine the truth? How can we be certain the Creator has always communicated through a chosen man? How can we be assured it is not simply the men of the society that invented this scenario to further their interests? It would seem we can never be assured of the truth as we cannot look back down the long hallways of time and seek it out. As they say, that ship has sailed.

But is it not natural to be a bit suspicious when we are told the Creator ignored all other humans that shared the planet with the chosen male? And is it not suggestive that such histories are configured so that there is no way to investigate them? They are lost in the depths of time with little to substantiate them. There were no news investigations, no television coverage, no photographs, and no independent coverage of any kind. Again, does it not raise some suspicion that these Revelations stories are always configured to enhance the position of men in their societies? And there is no way to confirm or dismiss them. We cannot access the lone man and no one else was present during the meeting so we can never ask questions.

A healthy skepticism, however, may be warranted when society proposes a Revelations story but does not provide convincing proof of its authenticity. Is it not perhaps wiser to retain some doubt of a story that cannot be authenticated, a story that one man could put forth without fear of exposure because no one else was there. A story that must be accepted on faith alone, a story that tells us the Creator always selects a man and never selects a woman? Can we ever search out the truth?

In many older societies of our world, men controlled the

avenues of communication. They did most of the story telling, writing, printing, and publishing. Accordingly, it would not be surprising if an account concerning the Creator and her Revelations was unintentionally crafted from the viewpoint of the men of the society. No one set out to mislead through an invented story of interface between the Creator and one male but perhaps the story simply evolved through repeated additions, modifications and editing over time to reach a final complex account. In this process the story probably progressed gradually from oral to written versions. It is not surprising that, in a male-dominated society, this process might have originated a account in which the Creator disclosed Revelations through one man.

But if the Creator actually wished to communicate to humanity, would she not just as likely communicate through one woman as through one man and is it not more likely, in any case, she would choose a wider communication channel than either of these? In other words, would she not act in accordance with her own wishes rather than those of men? And might not those wishes dictate that she communicate her Revelations to all rather than to just one male?

Yet several religions tell of how the Creator disclosed Revelations to one man in private. Perhaps there are reasons for this of which we are ignorant. For example, perhaps it was because these Revelations were communicated in a past time when women tended to stay in the background of tribal or community affairs. Could it be the Creator takes local customs into consideration when making her Revelations? Might she have a variety of issues to consider in making her decisions? And why would she not simply bypass the selection of one human to communicate to and simply pass her Revelations to directly to every human on the planet? Would that not be the simplest and most accurate method? Then again, she may have had reasons which will always be hidden from our simple inquiry.

Chapter 10

IN THE SCHOOL AND THE WORKPLACE, THE EFFORT OF HUMANS is often compared to accepted standards and/ or the efforts of others and rated accordingly. The measure of this rating is generally expressed as grades in school and in pay in the workplace. In either case, an effort that is judged exceptional may be rated as superior, an effort judged acceptable may be rated as average, and an effort judged to be wanting may be rated as failing. What if the same measure and rating were applied to the Creator's efforts to communicate her Revelations to humanity? Is it not exceedingly strange the Creator's efforts would probably be rated average at best and failing more likely?

Recall that at least thirty million people were spread about the Earth when the Creator is said by Judaism to have restricted the disclosure of her Revelations through just one man and that many times that number inhabited the Earth when the Creator is said by Christianity, Islam and Mormonism to have done the same. Yet, according to each of these religions, the Creator disclosed her Revelations through one selected man of the Earth, departed the field, and trusted the Revelations would somehow reach all other inhabitants of Earth with reasonable accuracy in a reasonable time period even though the inhabitants were scattered to the ends of the Earth and the existence of most was completely unknown to the selected man. Is it not curious the Creator would choose an error-prone communication system when an error-free one was readily available?

This same effort by a human would have been dismissed by a conscientious teacher or a diligent employer. They would have reviewed this performance and then asked, "Why did you not make more of an effort? You spoke through one man and vanished? You must be kidding. You know how fallible human communications are and yet you restricted the most important communication of all time on the planet Earth to one man? You should have kept at the task – you should have directly passed on your Revelations to as many humans as you could have reached - no, more than that, to every last human on Earth – they did not deserve less. Your performance was seriously lacking in thought, planning and effort – indeed, most disappointing!"

The Creator, of course, answers to no one but, for the sake of illustration, imagine she permitted an examination of her effort in a follow-up interview with an appointed inspector Ira. The interview might have gone something like this.

"Well, Creator, Tess tells me you disclosed your Revelations to those level-seven human critters out in the Milky Way galaxy. How did it go?"

"Great, I passed them on to a nice tall man."

"And ------ ?"

"And what?"

"Who else?"

"That was it – no one else."

Long silence and then, "Let me get this straight, you disclosed them through one man and took off?"

"You got it."

"These were Revelations for the ages and you just disclosed them through one man and split?"

"Well I don't think I'd put it quite that way, Ira, but, in a word, yes."

"I don't believe what I'm hearing – what were you thinking? What about humans on the opposite side of the Earth?"

"There's humans on the other side?"

"Come on!"

"Just kidding."

"Why did you convey your Revelations through only one man?"

"Well, I assume he'll pass them to the rest."

"Did you give him that assignment?"

"Not in so many words but ----"

"Does he know of the world? That it's round? That there are many continents?"

"It is? There are?"

"Stop that! I don't believe this – you have to know how unreliable human communications are – even if he does pass them on, after your Revelations have gone through ten hands by word-of-mouth, they will be mangled almost beyond recognition."

"Seriously?"

"Who's the Creator here? You know the communication skills at level seven and their short memory span. If it's not written down, forget it."

"I agree they're not too swift."

"Well, if your chosen man isn't even aware of people in remote locations of the Earth, how do you expect they will ever hear of your Revelations? It will take forever for your Revelations to reach around the Earth."

"What's your point?"

"What I'm saying is that your Revelations will probably never reach those far away and they'll be distorted beyond recognition if they do – why didn't you just reach out directly to all humans?"

"How could I do that?"

"I feel like I'm talking to an echo here – you're the Creator – think, there are a lot of ways you could have done it."

"You're right, I am the Creator and I'm starting to get a tad annoyed by this tough interrogation, Ira! You'd do well to remember who makes out the pay checks."

"Well, you just don't come up with important Revelations for Earth's humanity and then not make certain they reach each and every one of them in flawless mint condition – why not present them directly to each human – why select one man and, now that I think of it, how did you make your selection?"

"I just picked a nice, tall good looking fellow with a beard – he looked reliable."

"And his location - why that location and none other?"

"Like I told Chad, it had a nice dramatic biblical feel to it."

"Hoo-boy! Well, to be honest, I think you blew this but there will be other chances – when do you envision disclosing further Revelations?"

"I don't – I mean that's it – I'm out of here - I have no plans for further Revelations until these critters reach at least level eight and that's eons away and I'll be on the other side of the Universe by then - anyway, I think I covered everything – I do have other worlds in other galaxies, you know – it isn't easy being Creator and I don't like travel, it's not like I can spend all of my time speeding between clusters of galaxies."

"You mean you ignored humans in the ages before this and now plan on ignoring them again in the ages to follow – you're only going to make one visit to Earth and not return for ages?"

"I'm starting to get a bit annoyed again but yes, that's about the size of it - at least for the next million years."

"So out of the millions upon millions of people down the long corridors of time on this little Earth, you've only passed Revelations through one man on Earth and don't consider a follow-up any time soon?"

"You got it."

"Did you consider communicating your Revelations through a woman - I think Tess suggested that?"

"Get serious."

"How about conveying your Revelations through an assembly of humans so their combined recollections would confirm to others your visit?"

"Look Ira, I'm not that fond of them – especially in groups - I can take one at a time but ---- after all, they are level seven."

"The way you did it makes it seem like a secret – is that what you wanted?"

"No, my Revelations are for everyone."

"Then why not convey them to everyone?"

"Well, in principle I like humans - after all they're made in my image - but in person they make me itch."

"You don't have to do it in person."

"No?"

"Sure, you could blaze your Revelations across the sky like the northern lights."

"Really?"

"Or you could write them with arranged stars or write them on their Moon."

"You know you're pretty sharp, Ira – I'm glad I picked you as Inspector!"

"And you could communicate regularly – not just once and never before and never since – that must seem strange to humans."

"There're a little strange, themselves."

"Did you think about the fact that some humans won't believe the man you selected – that they will say it's all a fabrication?"

"I have a place reserved for them!"

"They couldn't say that if you had communicated to a large group – did you think of that?"

"I'm getting annoyed again – you don't seem to realize I'm the Creator. Anyway, what would you do, Mr know-it-all?"

"You're the Creator – you invented gravity, electromagnetism and the rest of those scientific thingies and probably have some more of them up your sleeve - you could appear simultaneously to each and every human on Earth and deliver your Revelations directly – you could write your Revelations in a great arc across the sky - why go through just one man? I thought the Creator was faultless and capable of any endeavor – how come you could create the Universe of hundreds of billions of galaxies and can't convey directly to a few million humans? Speaking of communication skills – are yours that limited?"

"I told you I don't like crowds, Ira – I get anxious."

"Then how about communicating to a small group of leaders that then each return and communicate the Revelations to their followers?

"I could have done that but I thought disclosure through one man had a certain flair to it - their reenactments of it in ages to come could be quite dramatic – imagine it portrayed up on a ceiling somewhere – I can see myself as a handsome older gentleman in a robe and with a long, flowing beard."

"Drama is more important than insuring accuracy of your Revelations?"

"Well, it doesn't sound so smart when you put it that way!"

"Why not at least communicate directly to different humans scattered over the Earth – some women and some men – your Revelations would then fan out to every human faster and more reliably?"

"I can do that?"

"OK, I can see you're not going to get serious - enough of this – your effort fell far short of expectations - I think you can guess my report to Chad is going to be ugly – not a pleasant afternoon of reading."

"And I don't think I need to remind you who the Creator is here, Ira - you can be replaced, you know – right now, I'm not a happy camper."

"Then why did you give me the assignment to be an Inspector?"

"Why, indeed!"

Chapter 11

IN ALL THE COMMUNICATIONS OF HER REVELATIONS, is it not curious the Creator has never left anything tangible for the later inspection of humans as proof of her visit? Without doubt, a tangible object from the hand of the Creator would be the most precious object in all of Earth. A great building would be constructed in a dedicated spot to house it. Sunlight would bathe it during each day and the soft glow of floodlights each night. It would be watched over by guards and by electronic surveillance. Everyone could admire it via the internet. Millions of humans from around the world would make pilgrimages to view it. It would be described endlessly in books and articles.

Replicas of it would set on tables or hang from walls in millions of homes. Smaller versions would be carried in purses and billfolds. It would serve as the fundamental foundation and proof for belief in the Creator and would essentially eliminate all doubt as to her existence, her creation of the Universe, and her bond to humanity. Who could doubt in the presence of such an object? It would be undeniable proof the Creator or her representative had visited the Earth. Is it not odd, therefore, that a tangible object from the Creator cannot be provided by any religion?

Stories about Moses generally place his life somewhere in the 13th to 16th centuries BCE. According to the Torah (e.g., the Book of

Deuteronomy), the Creator delivered to Moses two tablets of stone upon which were written ten Commandments. After Moses broke these tablets in anger at his people, the Creator commanded him to provide an ark and a second set of tablets to replace the first set. The Creator caused the ten Commandments to be written again on the second set and commanded Moses to place this set in the ark. No one today can produce either set of tablets nor the ark.

Little is known of the early life of Jesus but he was said to be born of a virgin. Around the age of 30, he was baptized by John the Baptist and began a ministry of teaching, healing, and miracle-working. After just a few years, however, opposition mounted against Jesus and he was crucified between 26 and 36 CE when Pontius Pilate was the Prefect of Judea. It was claimed that his body disappeared from his tomb and that he ascended into heaven. Although he never met Jesus, the apostle Paul was instrumental in spreading the Christian faith and his efforts ultimately led to the conversion of the Roman emperor Constantine in the early 4th century CE.

Beginning one night in 610 CE and continuing at different times until his death in 632 CE, Muhammad received the Revelations of the Qur'an from the angel Gabriel. For some time after his death, the Qur'an remained primarily an oral text until a scribe Zaid ibn Thabit was assigned the task of pulling it together. He is said to have stated with reference to Abu Bakr, who was the first in a line of Caliphs succeeding Muhammad, "By Allah, if he had ordered me to shift one of the mountains it would not have been harder for me than what he had ordered me concerning the collection of the Qur'an --- so I started locating the Qur'anic material and collecting it from parchments, scapula, leafstalks of date palms and from the memories of men."

In the early 1820's CE, with guidance by the angel Moroni, Joseph Smith found metallic plates buried in a stone box near Lake Ontario in the western portion of the state of New York. The plates bore writing in a language Joseph could not read. Joseph did not allow any one to view the plates because he said that would mean instant death. With the aid of a seer stone in the bottom of a hat, Joseph translated the writing and then returned the plates to their angelic guardian. No one today can produce the plates.

Thus, two of the world's religions are silent on the subject of tangible objects but two others state that tangible objects were provided directly to humanity from the Creator. Is it not regrettable that none of these objects are now available? The importance of such objects cannot be exaggerated. They would forever confirm the existence of the Creator and prove her concern for humanity. Even if the Creator made no further communications, her existence would be known and accepted by all.

Imagine the stone tablets had been rediscovered and placed on display. They would be the most valuable items in the world because they were associated with the hand of the Creator. Millions of people would travel each year to see them. They would provide indisputable evidence of the existence of the Creator and her disclosure of Revelations to Moses. Tiny portions of the stone would be subjected to scientific tests to determine its composition and origin. Attempts would be made to determine how the message was formed in the stone. The language of the message would be analyzed and dozens of scientific articles would soon follow.

Or imagine the golden plates are currently on view for everyone's inspection. They would be priceless because they came from the Creator. Thousands of people would come each day to view them. Millions would travel each year to see them. They would prove beyond doubt the Creator's conveyance of Revelations to humanity and, in particular, to Joseph Smith.

Because neither the tablets nor the plates can now be produced, it is not surprising that some rational people of sound judgment and sense doubt the stories behind them? Belief that an ordinary event occurred requires only moderate evidence whereas belief that an extraordinary event occurred requires positive evidence in support. When the extraordinary event concerns the Creator and is said to have occurred only once in recorded history, it is not surprising that many humans, in the absence of some reliable proof, will harbor some doubt about the event.

If told by a reliable source, most humans would believe it is raining even though they were not in a position to see or feel it. If told by a reliable source, humans would generally believe the Sun is eclipsed even though they were not able to personally witness it. If told by a reliable source, humans would believe that a dust storm is raging across the planet Mars even though they were not able to view it. If told by a reliable source, humans would typically believe that a distant star has gone nova even though they had no access to the instruments that showed it to be so. These are known natural occurrences and, accordingly, humans have no reason to doubt their existence if informed of them by someone they know to be reliable.

In contrast, even if told by a reliable source, it is unlikely that humans would believe that gold coins fell from the sky when such coins and the act cannot subsequently be produced. If told by a reliable source, it is unlikely that humans would believe that a river has reversed course and run up hill nor believe that grazing livestock suddenly ascended to the sky from their pasture. Even if told by a reliable source, it is doubtful humans would believe that a comet suddenly altered its direction so as to write a message across the sky when they did not personally observe this action. Such occurrences are more than just rare. As far as is known, they have never been observed in the history of our world and they defy the natural laws of the Universe. In the absence of absolute proof to the contrary, therefore, humans would typically doubt these occurrences even if told by someone they generally believe.

In the daily affairs of humans, skepticism would generally be the response to a statement that an extraordinary tangible object had existed but could not now be produced. The more extraordinary and unexpected the tangible object, the greater the skepticism. If a source were known to be reliable and claimed, for example, that hail had fallen on a stormy day, the source would generally be believed even if the event were not witnessed because the event would not have been an unusual one. If, however, the same source claimed that hail had jumped from the ground and sailed upward to the sky, this report would be met with great skepticism despite the source's reputation. This skepticism would be allayed only if considerable reliable evidence could be provided to confirm it. A common event would generally be believed even though not personally witnessed whereas an uncommon event would not.

Is it not logical, therefore, to doubt the Creator has ever provided tangible objects to humanity when they cannot now be produced? This is an extraordinary claim and, in the absence of the objects, it deserves to be doubted. Humanity is told the tablets disappeared when the temple was destroyed and we are told the plates were returned to the angel Moroni.

Humanity is asked, therefore, to believe extraordinary objects once existed but cannot now be placed in evidence. In their absence, it is not wise and prudent to harbor some doubt over the assertions? There is no merit in believing something remarkable in the absence of proof. To the contrary, it is reasonable to demand proof. There is nothing admirable about faith in the complete absence of evidence.

The Creator has been instrumental in forming us with the ability to assess facts and make judgments. Accordingly, she would know that doubt is a logical consequence when humans are told of the existence of extraordinary objects that are no longer available for inspection. After all, is it not reasonable humans would wish to see such remarkable objects for themselves? It is then not surprising we would withhold belief in an extraordinary claim in the absence of evidence. After all, a dishonest human can make such a claim as easily as an honest one.

Is it not therefore curious the Creator has never left anything tangible for the inspection of humans? Is this an unreasonable expectation? Having created humans, the Creator would understand their need for proof of unlikely events. The Creator would know that healthy skepticism is valuable in guiding humans down rewarding life paths and avoiding detours along unrewarding ones. Is it not therefore remarkable the Creator has never provided to humans a tangible artifact that would bear witness to her visit to Earth or to her Revelations? Would she not have considered the natural skepticism when a man states the Creator conveyed information through he and he alone? It would seem reasonable, therefore, that she would have left behind a tangible object as proof of her existence and her interest in humanity. Had she done this, there would now be absolute certainty of the Creator's existence and of her visit to Earth or the communication of her Revelations through her representatives. It may be understandable that the absence of a tangible object from the hand of the Creator permits some to harbor doubt.

Chapter 12

WHEN ONE MAN STATES HE AND HE ALONE
RECEIVED DISCLOSURES FROM THE CREATOR, it is natural to
be suspicious in the absence of some corroboration. Because she
created humans and is familiar with their inner workings, it would
seem certain the Creator would understand the skepticism humans
might have in response to the news she restricted her message to a
selected one of all the millions of people on Earth. And doubt is sure to
be increased when it is apparent the Revelations contain no knowledge
of the Universe that is not already known. This skepticism and doubt
would have been eliminated had the Creator disclosed something not
known to humanity at the time of her communication. When this
disclosure was confirmed in later times, humans would have had
incontrovertible evidence that the world had heard from her.

Being all-knowing and all-wise, would she not have thought to
include an unknown truth, an unknown law or principle to verify her
presence? When, in the long history of humanity, this disclosed
knowledge was later shown to be true, there would have been no doubt
of her visit. Imagine, for example, she had disclosed to Moses that it's
the Earth that moves about the Sun and that Michelangelo had included
that knowledge when he painted the ceiling of the Sistine Chapel
between the years of 1508 to 1512. When it was mathematically
demonstrated in 1543 by Copernicus that the Earth orbited the Sun, all

of Earth's humans would have known it was indeed the Creator who had spoken to Moses.

Or consider that the Creator had disclosed through Jesus that the Earth was a huge substantially-round ball and not just the plain bordered by and defined by the horizon. When observations later began to corroborate this, humans would have had proof the Creator spoke through Jesus. Yet again, imagine the angel Gabriel had disclosed to Muhammad that the planets moved about the Sun in elliptical orbits. When Johannes Kepler worked out the mathematics of planetary motion in 1609, all would have been certain we heard from the Creator. And what if the Creator had informed Joseph Smith through the angel Moroni of the presence of the outer planet Neptune? When humans positively observed Neptune in 1846, they would have known without doubt that the Creator had sent word to humans.

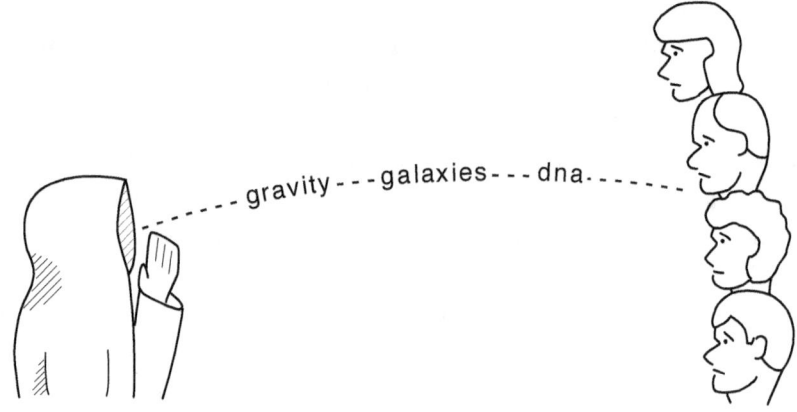

Figure 13
"an unknown truth, an unknown law or principle to verify her presence"

Imagine the Creator had mentioned to Moses that moons circle the planet Jupiter, that blood circulates through the body and is pumped by the heart, that there are microorganisms and someday bacteria would be viewed through a microscope, that every particle of matter attracts every other particle with a force along the straight line joining them and this force is directly proportional to their masses while inversely proportional to the square of the distance between them, that all forms of life could be ranked and classified on the basis of shared physical characteristics, or that oxygen is an element involved in combustion and respiration. When this knowledge was respectively confirmed by Galileo in 1610, by William Harvey in 1628, by Anton van Leeuwenhock in 1674, by Isaac Newton in 1679, by Carl Linnaeus in 1735, and by Joseph Priestly in the 1770's, humanity would have known for certain the Creator had spoken.

Figure 14
"it's the Earth that moves"

Consider if the Creator had disclosed through Jesus the simple principles of a telescope, that plants use photosynthesis to convert the energy of sunlight into chemical energy, that lightning is an electrical discharge, that billions of stars are arranged to form the Milky Way galaxy, that light is an energy wave, or that certain chemicals could be used as anesthetics to facilitate surgery without pain. When this knowledge was respectively discovered by inventors in the time of Galileo, by Jan Ingenhousz in the 1770's, by Benjamin Franklin in 1751, by William Herschel around 1800, by Thomas Young in 1802, and by various scientists in the 1840's, we would have known the Creator had been with us.

Imagine yet again the Creator had revealed to Muhammad that

people could be protected from disease by the process of vaccination, that atoms in elements could combine in numerous ways to form molecules, that energy can neither be created nor destroyed but can only be transformed from one state to another, that sound frequency is altered when its source is moving and of the existence of electromagnetism. When these principles were confirmed by Edward Jenner in 1796, Amedeo Avogardo in 1811, Julius von Mayer and William Thomson in the 1840s, by Christian Doppler in 1842, and by James Clerk Maxwell in the mid-1860's, all doubt would have been erased.

And what if the Creator had mentioned to Joseph Smith that animal cells divide in a process called "mitosis" and that plant cells do likewise, that X-rays could be used to view bodily structures, that the age of rocks could be dated via the decay of radioactivity, that Neanderthals had shared the planet Earth with Homo Sapiens until approximately 30,000 years ago, that light is bent by gravity, and that the Universe contains billions of galaxies moving away from us at a rate that increases with their distance from our galaxy? When these facts were respectively discovered, by Walther Flemming and Eduard Strasburger in the late 1880's, by Wilhelm Roentgen in 1895, by Bertram Boltwood in 1907, by various investigators beginning in the 19th century, and by Albert Einstein and Edwin Hubble in the early part of the 20th century, we would have known we had been in the presence of the Creator.

And imagine the Creator had disclosed to Moses that all elements consist of particles called atoms, that laws of thermodynamics define temperature, energy, and entropy and describe the transfer of energy as heat and work, that light has a constant speed whether radiating towards or away from an observer, that our Australopithecus ancestors lived 3.5 millions years ago, and that mitochondria perform metabolism in living cells. When this knowledge was respectively discovered by John Dalton in 1808, by Benjamin Thompson, Sadi Carnot, Rudolf Clausius, and William Thomson between 1797 and 1873, by Albert Michelson and Edward

Morley in 1887, by Mary Leaky in 1978, and by various investigators in the 20th century, it would have been certain we had heard from the Creator.

What if the Creator had disclosed through Jesus that life evolves through a process of variation and selection, that a limited number of chemical elements form the Universe, the principles of radioactivity, that negatively-charged particles are part of all atoms, that the terrors of malaria could be eliminated, or that evolution developed several different life forms in the early times of the Earth and left proof of them in the Burgess shale? When this knowledge was respectively confirmed by Charles Darwin in 1858, by Dmitri Mendeleev in 1869, by Marie and Pierre Curie in the 1890's, by J. J. Thomson in 1897, by Ronald Ross also in 1897 and by others subsequently, and by Charles Walcott in 1909, we would have known they followed in the footsteps of the Creator.

Consider that the Creator had revealed through Muhammad that the DNA molecule is the chemical basis stored in the chromosomes that carries genetic information between generations, that mass and energy are manifestations of the same thing and energy of a mass equals the product of that mass and the square of the velocity of light, that there are billions of galaxies in the Universe, that penicillin could kill bacteria, that there had been hundreds of different families of dinosaurs and that sixty five million years ago an asteroid impacted Earth and killed the last of them, and that "black holes" exist at the centers of galaxies with mass sufficient to absorb light. When these principles were confirmed by James Watson and Francis Crick in 1953, by Albert Einstein and Edwin Hubble in the early part of the 20th century, by Alexander Fleming in 1928, by Walter Alvarez in the 20th century, and by scientists in the latter part of the 20th century, we would have known of the hand of the Creator.

Finally, what if the Creator had informed Joseph Smith that electrons travel in restricted energy orbits, that continents are carried about Earth on huge plates, that DNA triplets encode twenty amino acids to build the proteins of each human, that an early human species

Australopithecus afarensis populated the north-western African highlands, or that there are planets about other stars of the Universe? When this knowledge was respectively discovered by Neils Bohr in 1913, by various scientists in the 1960's, by Marshall Nirenberg's team in the 1960's, by Donald Johanson in 1974, and by observations in the 21st century, we would have known this knowledge could only have come from the Creator.

Yet not one of these facts, truths, laws or principles has ever been disclosed by the Creator prior to their independent discovery by humanity. Neither Moses, Jesus, Muhammad, Joseph Smith nor any other human has ever told of receiving from the Creator knowledge not already discovered by humanity. And yet it would have been so easy for the Creator to have revealed something unknown to thereby assure humans they had indeed heard from her. One mention that the Earth was round, that it revolved about the Sun, that Moons orbited the planet Jupiter, that light has a constant speed whether radiating towards or away from an observer, or that the Universe is expanding and all would have known we had been in the Creator's presence.

One hint that life evolves through a process of variation and selection and we would have been certain. One mention that the Sun was one of billions of stars forming the Milky Way galaxy. One word about electrons, or of gravity and we would have known. One clue about bacteria, dinosaurs or an early human family such as Australopithecus afarensis. One whisper of DNA, of Neanderthals, or the Burgess Shale.

The earliest evidence of writing dates to about 3000 BCE so we do not know what knowledge humans might have received from the Creator prior to that time. We only know that since the advent of writing, no piece of knowledge has ever been received prior to the time this knowledge was independently discovered by the human race. The Creator never mentioned to Moses that the Earth is round. The Creator never thought to have Jesus disclose the Earth revolves about the Sun. The angel Gabriel never informed Muhammad that the planets moved about the Sun in elliptical orbits under the control of gravity. And there was nothing in the golden plates to tell Joseph Smith about electrons, bacteria, electromagnetism, the genetic code or the billions of galaxies that march out to the edges of the Observable Universe. Because the Creator's silence on such subjects has been constant and unvarying over the ages, humans have no independent assurance they have ever heard from the Creator. Of these subjects the Creator's silence has been constant down the corridors of recorded history. And yet still it has long been told that the Creator communicated to us through Moses, Jesus, Muhammad and Joseph Smith.

Chapter 13

THE VARIOUS DISCLOSURES OF THE CREATOR HAVE ONE THING IN COMMON – they never reveal anything to humans not known at the time. If they had, there would now be no doubt they came from the Creator. Suppose, for example, the Creator had told Moses "your children will learn their home is in a wheel of stars surrounded by billions of other wheels". When the existence of galaxies was determined in the early part of the 20th century, there would have been no doubt it was the Creator who had spoken to Moses - who else would have known of galaxies? Or imagine the Creator told Moses "thou shall learn from the code of life stored in each of your cells". Because only the Creator could have made this reference to DNA and to the cells that contain it, succeeding generations would have been assured that the Creator had spoken to Moses.

In fact, however, there has never been a reported Revelation that recounts anything not known to humanity at the time. Among skeptics, this leads to the suspicion that the Revelations were originated not by the Creator but by humans. Surely the Creator would have revealed something not already known on Earth. The Creator's public relations manager Pam is concerned to avoid this very situation.

"Quick, what is it Pam? I'm busy – I'm packing for my trip to a group of galaxies out in the Coma supercluster – sometimes I regret creating so many of these galaxies when one or two or twenty would have been enough – I always say, "when you've seen one galaxy, you've pretty much seen them all. Right?."

"Yes, we know you always say that, Creator – catchy saying. Well, it's just that doubts have been coming in lately from people on Earth – some of them don't believe you really communicated your Revelations to one of their kind in your last tour through the Cosmos with Tess."

"Earth – Earth – doesn't ring a bell."

"The one with people, mankind, humanity – that tribe."

"The ones with the arms arranged like a sunburst about the head and that hop on one leg?"

"No."

"The ones with front and back arms, heads at each end, and an endless track instead of legs?"

"No – the ones that stand on two legs and wave two arms around a head."

"Oh, yeah! Tess and I visited them. That was one of my early designs – didn't really work out too well – I remember - OK, what's their problem now?"

"They say there is no way to confirm the Revelations came from the Creator because there is nothing in them they didn't already know. I've tried to warn you before that this sort of thing trashes your reputation in the Universe."

"Holy galaxy, that old argument again – I'm so tired of it. Anyway, I already went over all that with Ira."

"Yes, I know - I read his report. He was just concerned that you only spoke to one man. But I was shocked to learn you didn't disclose anything these humans hadn't already worked out on their own. We've gone over this before. One man says he and he alone spoke with the Creator – naturally, people are going to be skeptical – he could just be another nutcase – but if he reports something that becomes confirmed only in future generations then everyone will know it had to have been the Creator. It's all about image."

"It makes me feel cheap to have to prove myself."

"Still - why not? For example, why didn't you just tell them the Earth is round? They didn't find that out for centuries after your visit – that would have shown later generations beyond doubt you were the Creator."

"The Earth is round?"

"Come on!"

"I always get a kick out of that little joke."

"Look, they have a point. When you were last there, they still assumed the Earth was flat – why didn't you let them know otherwise – they would have been infinitely impressed. Later generations would have been assured you were the Creator."

"I like to just stick to giving them rules – that keeps them quiet and orderly and maintains discipline."

"But why not divulge something unknown to them. That would provide absolute proof in years to follow that they heard from the Creator. It doesn't cost you anything."

"Such as? What would have been needed? You know the level these people are at, Pam?"

"I know - level seven – it wouldn't have taken much."

"I'll say – level seven – boy, I haven't seen any of those for years! Oh, wait – I remember now, things were slow back then so I tried a few level sevens. Didn't really pan out so after our last reorganization I decided no more of those."

"You could have told them their Earth revolves about their Sun – when they got to the point later where they worked that out, they would never, ever again doubt they had heard from the Creator."

"If I do all the heavy lifting, what's left for them?"

"That's not the point, they would still have had to work it all out but when they got there, they would have said, "Wow, that really was the Creator speaking back then!""

"I don't know – it sort of cheapens the whole thing – you know what I mean, Pam? I'm the Creator yet I have to prove myself each time."

"But what's the point of communicating your Revelations if

you do it in a way that anyone could have? How are they to know you're not a fake or a rumor or a hustle or a scam? You're just in and out, speak to one man with a beard, and then they have no proof when you're gone."

"Well, I never thought of it that way – you're pretty sharp Pam – I'll remember this at review time."

"As a Creator, you do great things, galaxies, clusters and all that, but when it comes to critters like these humans, you seem a bit dense if you get my drift."

"Easy, Pam – I can get mighty testy as you know."

"OK, but back to your Revelations to humans on Earth - it didn't have to be much – just reciting the continents would have done it – they would have been convinced when they discovered them later right where you said they were."

"Problem with that one is I'm not too up on them myself. I can never keep South America and Africa straight – they always look the same to me."

"My staff could have filled you in before the visit."

"But other than that and the Earth revolving about their little Sun and being round, I don't know what I could I have included in my Revelations to convince them."

"Are you kidding? Things like elements, planetary motion, the solar system, and continental drift. Like gravity, electrodynamics, absolute zero, evolution, animal and plant cells, and blood circulation. They knew of none of these. Things like galaxies, black holes, the atom, relativity, speed of light, the big bang, and the expanding Universe. Like how to use anesthesia to alleviate pain and how to use X-rays to view body structures? The mention of any one of these would have done it."

"I'm still not sure about that evolution thingy myself, Pam."

"Oh, come on! We've been all through the evolution thingy!"

"I know, but in the quiet of the night, I still wonder sometimes."

"Keep it down – I don't want to get the staff going on that one again. Anyway, imagine you had told them about DNA? Holy smoke! When they worked that baby out later, they would have believed in you beyond a shadow of doubt!"

"Yeah – I like that one myself – kind of outdid myself there – it took some steam out of me – haven't really been up to that level now for a long time – slowing down a bit, I guess."

"The point is you just gave them some rules to keep them quiet and orderly and maintain discipline. But there was nothing there that anyone of the time couldn't have come up with so how were they to know if it was the Creator rather than an impostor?"

"I sort of see your point here."

"While we're at it, why didn't you eliminate at least one of their diseases or at least warn them just for the sake of public relations?"

"Warn them about what?"

"Well, how about the plague, cholera, yellow fever, typhus, polio, or small pox – anyone of those would have helped. The point is the people on Earth have to take it on faith that you visited because you didn't reveal anything they didn't already know."

"So?"

"Now that I think of it, how come you never even inadvertently disclose something they don't already know? You'd think something would slip through once in awhile. Like you could have mentioned that clocks run slower and light rays bend in strong gravitational fields."

"Go on! They do?"

"OK - Tess warned me how you play ignorant when it suits you. We're on to your little tricks."

"Pam, I don't think I like where this is going. Maybe you'd better sum things up before you and Ira get booted into the Sculptor void."

"OK, here it is - is it not puzzling you failed to throw in one fact they didn't already know and only spoke to one man? How could they be assured you were the Creator? Isn't this just what a faker would have done? You forced them to take it on faith but how were they to know it wasn't all a sham? This sort of thing wrecks your reputation and its my job to keep your image shiny and squeaky clean. To make sure this doesn't happen again, my staff has come up with a list of things you can disclose next time to prove yourself beyond all doubt. Shall I schedule a meeting to go over them?"

"Holy Creator, Pam! Will you look at the time! I have to run to the other side of the Universe! Call me sometime and we'll do lunch."

Chapter 14

IN OUR MILKY WAY GALAXY, THE ORION SPUR IS SPACED OUTWARD some 26,000 light years from the center of the galaxy. It begins at the Sagittarius arm and angles outward 10,000 light years to where it merges into the Perseus arm. Although a minor arm of the galaxy, the density of stars in the Orion spur is such that there are probably 15,000 stars within 100 light years of Earth and nearly 2 million within 500 light years. Is it not strange the Creator who formed this immensity of star structures would be content with delivering to one man two sets of stone tablets that have since disappeared? Is that a credible effort by the Creator of such massive assemblies of stars? Perhaps our small planet is so insignificant in comparison to the rest of the Universe that she felt no need for more of an effort?

Our local Virgo supercluster has 200 trillion stars in 2500 large galaxies distributed across over 100 million light years. To reach Earth, the Creator would have had to first ramble through the Virgo supercluster to find the Local Group of galaxies and then move on to our Milky Way galaxy. Is it likely the Creator that built this supercluster would restrict the travels of her son to the lands bordering the eastern end of the Mediterranean Sea when she wished to communicate to humanity? Would she really have been content with this effort, knowing how long it would take for her Revelations to spread to some 170 million people about the Earth?

Although our Milky Way galaxy contains between 200 and 400 billion stars arranged across some 100,000 light years of space, it is dwarfed by the Shapley supercluster which lies 500 million light-years away and is perhaps the most massive association of galaxies in the Observable Universe. Is it not curious the Creator that assembled the massive Shapley would only convey Revelations to one man during his lifetime and ignore all other humans? In Muhammad's time there were at least 200 million people on Earth yet the angel Gabriel would make no attempt to reach them but speak to just one?

Almost twice as far away as the Shapley supercluster, the Horologium supercluster is not as dense but its galaxy clusters are scattered across half a billion light years so that, in area, it is one of the largest known superclusters. Thousands of galaxy groups are scattered across 120 million light years and contain almost ten thousand times as many stars as our galaxy. Is it not peculiar the Creator that formed the Shapley supercluster would limit her communications to directing one man to some buried golden plates? When she wished to bring Revelations to humanity, is it likely the angel Moroni would travel past the millions of stars in the Orion spur to find the Sun and then travel out to Earth and, after this great journey, speak to just one man?

The Creator arranged over 200 billion stars to form the spirals of the Milky Way galaxy yet her Revelations to humans were limited to two stone tablets? She packed 200 trillion stars into the Virgo supercluster but decided her son Jesus need not travel beyond the lands at the eastern end of the Mediterranean Sea? She scattered thousands of galaxy groups across 120 million light years to form the Shapley supercluster yet felt it sufficient for the angel Gabriel to contact just one man? She arranged the Horologium supercluster across a space so vast that light requires 500 million years to cross it but settled for burying some metallic plates?

One man who says the Creator twice met him, and him alone, and provided Revelations on stone tablets that no one can now examine. A second man who was said to be the son of the Creator and to have arisen from the dead. A third man who received Revelations from the angel Gabriel. And a fourth man who found buried plates with a message from the Creator written in a language the man could not read but no one can now examine because he gave them back. This is it in the 5000 years of recorded history? It would seem the Creator could do better in a single day. If the Creator saw fit to leave her Revelations in a tangible form such as stone tablets or metallic plates, surely she would have arranged they would remain for the ages for the benefit and inspection of humanity. Yet we are told these tangible objects existed at one time but have now disappeared. Does the Creator really play such games?

These tenuous disclosures are all the Creator of the Universe can muster when looking after humanity over thousands upon thousands of years? The Creator that could construct our solar system with its huge Sun and its circling planets would ask humanity to believe in simple stone tablets that cannot now be produced? The Creator that built millions upon millions of vast galaxies and set them hurtling into the black voids of the Universe would communicate her disclosures by sending one son and then taking him back never to be seen before or since? The Creator whose billions of stars project light over staggering distances would restrict her Revelations to one man in a cave? The Creator that initiated the nuclear fusion processes of our Sun to convert mass into prodigious amounts of energy would convey her Revelations by burying them, guiding only one man to find them, and then taking them back again?

Imagine a first man comes up to you and says the Creator gave him Revelations written on stone tablets. You might respond, "Really! I would love to see them!" The man replies, "You can't because I broke one set of them and a second set was lost." Most rational people would then say something like, "Would you look at the time? I've got to run!" Or imagine a second man approaches you on the street and says he is the son of the Creator and has risen from the dead. A logical person would probably say, "That's fascinating – wish I could hear more but I'm late for a meeting." Imagine yet a third man approaches and says he has been hearing messages from the Creator. A cynical person would probably say, "I'd like to hear some – are you recording them?" Finally, imagine a fourth man confides in you that he has deciphered buried plates which an angel led him to. Some people might hopefully ask, "That's wonderful – can I see them?" When told the angel took the plates back, most might say, "Isn't that the way it goes? Good things never last."

These are the reactions most people would surely have today yet have faith in similar stories about Moses, Jesus, Muhammad and Joseph Smith because they come from a distant, poorly-recorded past. Is it not more likely the Creator would simply appear before all of humanity and present her Revelations in a form that leaves no room for questions or doubt? Would not the Creator take caution and present her Revelations in a manner that all would be certain of their origin? Surely one who devised electromagnetism could easily originate signals to reach each human on Earth no matter where they might abide. Is it then not highly unlikely a Creator so powerful as this would communicate Revelations to humanity by restricting them to just four men over the 5000 years of recorded history?

It must be recalled we are speaking here of the Creator who formed scores of different types of atoms – the building blocks of the Universe, developed gravity and electromagnetism, flung billions of galaxies across the Universe, populated each of these galaxies with billions of stars, generated nuclear fusion in each star to convert hydrogen to helium, neutrons and massive quantities of energy and, finally, circled the stars with planets. Would a Creator capable of such incredible accomplishments then shy away and convey her Revelations to humanity with stone tablets and metallic plates that cannot later be produced or with a story of virgin birth or with a story of intermittent angelic communication through one man? Would the Creator really prefer to leave her disclosures shrouded by doubt rather than enshrined in fact? Is it credible the Creator prefers that her existence must be taken on faith alone rather than simply established by fact?

There has been a huge expansion of human knowledge of the Observable Universe in recent times and it is wise and prudent to consider that knowledge when judging claims that the Creator restricted her communications through one man. The Observable Universe is so vast that were the Creator to voyage from its edge to reach our Milky Way galaxy, it would be equivalent to traveling 20 miles to reach an object the size of a quarter. Even our galaxy's size is so huge that if the Creator then traveled through it to find our solar system, it would be equivalent to traveling through the North American continent to find another object the size of a quarter.

Is it not strange the Creator would make such an incredible journey to reach one tiny planet far out on one spiral arm of a one lonely galaxy among billions of others, then disclose Revelations through one man and leave forever? It has never been a teaching of any religion that the Creator repeated this incredible journey. It has never been a teaching of any religion that the Creator communicated before or after this lone communication. Knowing she would only voyage once and never return, would not the Creator wish to disclose to most, if not all, humans and certainly to more than one man before leaving forever? This is it in over 5000 years?

Chapter 15

IF FRED CAME TO YOUR OFFICE TOMORROW MORNING AND CLAIMED THE CREATOR gave him Commandments inscribed into stone tablets which were now lost, would you assume the story to be true? Not likely. If Fred came to your table later while you were having lunch and told you he was the son of the Creator who had sent him to Earth with a message for humanity, would you trust his story? You know you would not.

If Fred came to your house that evening and told you the Creator had spoken to him as he was meditating in a cave, would you have faith in his story? Probably not. If Fred came up to you on the street tomorrow and told you the Creator had guided him to buried gold plates that carried a message in a strange language, provided special spectacles so he could translate the message, and subsequently retrieved the plates so that they were no longer accessible, would you believe him? Of course, you wouldn't.

Yet the same stories told hundreds and even thousands of years ago are believed by millions of people around the world today. Perhaps because they are viewed through the clouded, enhancing and distancing glass of time. The vision of Moses in antiquity seems inspiring – Fred in your office tomorrow lacks the same gravity. Jesus on the cross is unforgettable – Fred at lunch may be something less. Muhammad in his cave with the angel Gabriel is a noble vision – Fred at your door may seem an intrusion. Through the mists of time, the story of finding buried gold plates may be intriguing – on the street tomorrow, the story is more likely to be brushed aside as fantastic.

Figure 15
"imagine a press conference today with a modern Moses"

Imagine a press conference today with a modern Moses named Fred who describes how he recently received a message from the Creator on stone tablets. He would be illuminated by flashes from dozens of cameras as he stood behind a bank of microphones and tried to answer questions called out from all sides. It would be one of the most viewed interviews of recent times. It would be all over the internet, shown on every evening news channel, and there would be hundreds of written accounts in newspapers and magazines in the following weeks and months.

Dozens of news organizations would ask permission to examine the tablets. Scientists would request samples of the stone so that its age and composition could be ascertained and analyzed. The text would be magnified and studied to determine how it had been formed in the stone. Language experts would examine the sentence structure in great detail.

In interviews, Fred would be asked to describe the Creator – "what did she look like? - what was she wearing? – did she have a human form?" "What language did they converse in? How did she make contact with him? "How did she arrive?" "How long did they converse?" "How did she leave?" "Does he know why she communicated through him and only him?" "When and where did the visit take place?" "Did she say there would be further visits in the future?" Everyone would watch the interviews and study his every word. Opinion polls would find that many of the public doubted him and found errors in his story. But others would probably be convinced and would ecstatically receive this modern Moses.

If the scientific tests determined the stone's origin was not of this Earth and the formation of the inscription in the stone was also different from anything previously known, the intensity of the press and the interest of humanity would grow enormously. If they instead showed the stones to be a crude forgery, the story would quickly drop from the world's interest and would soon be forgotten. The words and reputation of the modern Moses would be quickly overwhelmed by the negative turn of events. They could not long survive the intense scrutiny of modern communications.

Because events now flash about the Earth in minutes and can be examined in detail by nearly everyone, it is unlikely that a new Revelation story will ever appear unless it is authentic and proven to be so. And it was probably also difficult for one to become established early on in the 5000 years of recorded history. At that time, the story would have remained local and could not have reached enough people to have become established. Around the times of Moses, Christ and Muhammad, however, a Revelation story could reach enough people to become established as a part of their belief system but would not have been subjected to the intense study, analysis, and questioning of the interconnected world of today.

It could prosper in a social environment that allowed it to slowly spread across the lands at the eastern end of the Mediterranean Sea but did not expose it to an aggressive intellectual study or a penetrating scientific examination. The Torah does not report that anyone questioned Moses or that he discussed the events with anyone. The story is just reported as factual with no details to support it. In marked contrast to the Revelation stories of Moses, Jesus and Muhammad, that of Joseph Smith originated in the early nineteenth century. This must surely be as late in recorded history as a new Revelation story could occur. It would have been reported and studied in much greater detail had it come along one hundred years later.

It is notable, however, that each of these Revelation stories occurred only once and was never repeated. Is it not curious the Creator would only convey Revelations once in the 5000 years of recorded history? Is it not more likely she would communicate early, late and often? Surely she would be as likely to communicate now as two thousand years ago.

What would prevent her from communicating over and over? And why would she have ignored millions of people that lived in the earlier portion of recorded history? If it was important to convey later when she did, was it not equally important to communicate earlier? And would she not be as likely to communicate in our times? Would she not, in fact, have been a constant visitor throughout recorded history? It seems unlikely she would communicate to humans once and never before nor since.

If, however, the Revelation stories were initiated by humanity and completed by repetition and enhancement over time and generations, it is understandable that they would have been created only once and only in a particular portion of recorded history. Earlier in that history there would not have been sufficient disclosure to establish the stories and in our times unsubstantiated miraculous stories could not have survived the intense focus of the internet. Only in the times between these eras was it possible to originate a Creation story that would then gather enough authority and weight to survive the questioning of later generations. This story would most likely describe a single visit by the Creator because humans would find it difficult to invent, describe and authenticate a series of visits. There would have been no incentive to add to the original visit and stories of additional visits might have stretched human credulity beyond the breaking point.

If the Creator has an interest in the welfare of humanity, however, is it not passing strange she would look in on us only once in the 5000 years of recorded history? Is it not odd she would introduce doubt over the authenticity of her single visit by speaking to just one man in private, not mentioning anything to him of the millions of people around the Earth of which he was ignorant, and not informing him of the locations around the Earth of these millions? Are we to believe the Creator that formed the mighty Universe and the laws that govern it would have been so casual and indifferent? Is it not perhaps informative she has been silent and absent since the advent of modern communications which expose actions and events around the Earth to everyone's immediate attention?

If the Creator cares about humanity, would she not have ranked the health and well being of humans at least as important as her Commandments? Would not awareness of the dangers of the bubonic plague have been of equal value to humanity as the admonition to not kill? Would not caution of the cause and prevention of small pox have been at least as valuable as the command to not commit adultery? Would not information of the cause and prevention of malaria been more useful to humanity than the injunction to not steal which all communities would eventually originate on their own? Would not a warning of the cause and the dangers of AIDS have been at least as valuable as the instruction to not bear false witness? The latter are social mores humanity later worked out for themselves. The former are illnesses that imposed enormous suffering before humanity solved them. Dear Reader, was not relief from these diseases worth at least a second visit by the Creator?

Chapter 16

OVER THE AGES, A GREAT DEAL OF ATTENTION has been directed to the Commandments recited in the biblical books Exodus and Deuteronomy. We are told they were inscribed on two stone tablets by the finger of the Creator and given to Moses to thereby provide a guide to moral standards. It might be of interest to compare these Commandments to other ethical standards that would seem important in guiding people to behaviors that enhance humanity's life on this little planet.

The first four Commandments can be generally ignored as they do not address moral or ethical behavior but, rather, are internal "housekeeping rules" typical of those that might be directed to members of a club or a social order. Injunctions to not have any other gods, to not keep idols, or to not take the Creator's name in vain have nothing to do with admirable social and moral behavior. They don't cause one to be a better neighbor or a more valuable member of society. Instead, they seem designed to assure and maintain respect for the organization. In a similar vein, a reminder to remember the sabbath day can be compared to club notes such as "meetings are every other Tuesday", "Mondays are reserved for members only", and "bingo every other Wednesday".

And a reminder to honor your mother and father can't be faulted but is it really necessary to prescribe what is generally a heartfelt emotion? Must we make it a rule to like your best friend, to admire your teacher, to love your dog or to appreciate nature? Is it necessary to demand feelings of love and respect? Would it not be sufficient, and indeed better, to allow such feelings and emotions to take place naturally?

But then we come to the "shall nots". You shall not kill or steal. One would think these rules must be an inherent part of any established society. And while not necessarily illegal most people would naturally frown on adultery and lying. As to the final Commandment to not covet your neighbor's property, this Commandment is not only impossible to observe but it is not necessarily one that would improve society.

This Commandment seems to confuse the natural desire to earn for oneself something similar to that possessed by a neighbor with the destructive desire to actually take that possessed by the neighbor. After all, covet does not mean "to take" but rather "to desire" or "to wish for". Wishing you could have a luxury automobile similar to your neighbor's automobile is a natural feeling. It is the response to coveting that needs to be controlled. That is, you work to obtain that automobile - you don't steal it.

In any case, the world has many moral issues at least as important as those expressed in the Commandments of Exodus and Deuteronomy. We may learn of just a few by listening in on the following conversation between the Creator and her research director Ron.

"Thank you for seeing me on short notice, Madame."

"Whoa! How many times have I told you to knock first – you scared the celestial pudding out of me!"

"Sorry about that Madame – it's just that this is so urgent."

"And what have I told you over and over about addressing me? We need to keep some semblance of respect and decorum in this Universe."

"What? – oh yeah, I forgot. Thank you for seeing me, Madame Creator".

"OK, what is it this time, Ron? We have to keep this short – I'm creating a new beetle species for a planet in a galaxy in the Fornax Cluster and that's more than 40 million light years away – I'll be lucky to not miss lunch.

"You really like beetles don't you, Madame Creator? There's more than 350,000 recorded species on Earth alone."

"They don't talk back, they don't take up much room, and they look neat pinned to the wall in my study – what's not to like? But what's this Earth thingy, Ron?"

"I'm getting to that now Madam Creator. My staff has been doing deep research in which we collect all the available Universe records on catastrophes in various galaxies. We then look through the plans on the big Universe computer to find similar events scheduled for the future. We've had good luck using this process to predict and prevent similar tragedies that may occur in later societies. I've just come from a meeting in which the staff summarized the results of some of their investigations. One of them concerns explorations in which travelers from one planetary continent will come in contact with inhabitants of other continents that have been isolated for millions of years. The consequences of this sort of thing can be immense – we need to prepare now to prevent these on some of your younger planets. My staff pulled out a few cases for closer inspection and found some concerning this little planet called Earth that call for immediate attention."

"Never heard of that planet."

"Of course you have – you visited it a couple of eons ago with Tess – it's in a galaxy called the Milky Way and the level seven inhabitants are called humans."

"Isn't that just the cutesy name you'd expect from level seven? I remember them now – how could I forget? And I also recall going over that visit with Ira and Pam. These humans seem to need a lot of stroking - what's their problem now? "

"We anticipate an explorer named Columbus will sail across one of Earth's oceans to find a continent and accidentally discover a different one on the way. As he approaches an off-shore island he will see lights ahead and decide to halt his ship for the evening. We anticipate the island will be visible at first light in the morning and Columbus will go ashore."

"Good luck to him. At level seven, I'm surprised he can find anything."

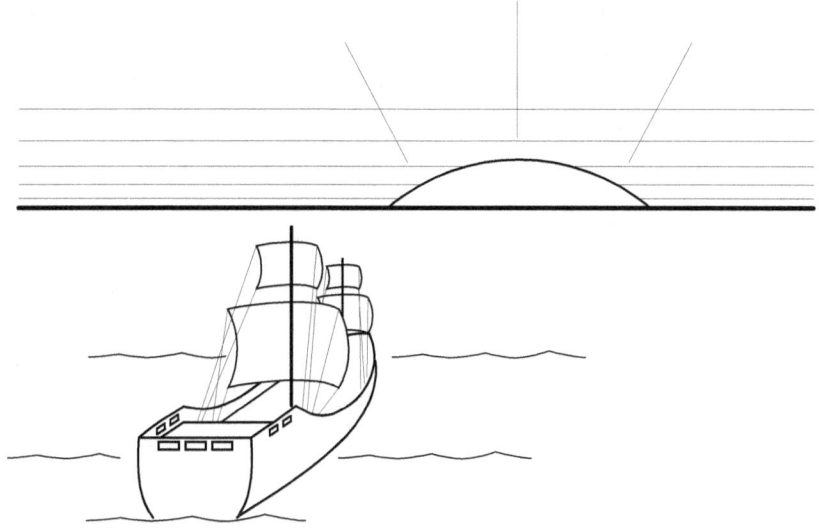

Figure 16
"the island will be visible at first light in the morning"

"Yes, but here is the problem. The island's inhabitants have never been exposed to a deadly disease known on Earth as small pox. Because of the way you set up DNA, they have no resistance to it and it will literally wipe them out. The suffering will be terrible. We will have to do something before this sunrise at the island to prevent a catastrophe. Uhh – Madame Creator?"

"Hmm? Oh, sorry, Ron. But this reminds me how fond I am of my Commandments and I was just thinking of a good one for this situation. Look, just tell them not to take my name in vain.

"Yes I'm sure that will come in handy Madame Creator but back to the problem at hand. Columbus has no way of knowing the consequences but if we let his men step ashore in the morning, thousands upon thousands of the local inhabitants will die terrible deaths from small pox in just the next few years. Their innocent children will be especially vulnerable."

"Just tell them to keep the Sabbath and all will be well – now let's move on – what else you got."

"With all due respect, Madame Creator, they'll still die whether the Sabbath is kept or not."

"Can't hurt though. And don't forget – tell them not to take my name in vain."

"I don't believe this – if we let this happen, all will be lost for these people. We know the terrible consequences – they have no clue. Can't we do something to prevent the suffering and death that will result?

"I wonder what's for lunch in the Fornax Cluster?"

"Your lunch is more important than preventing a terrible disaster?"

"It's just that you're so serious Ron, and thinking always makes me hungry."

"OK, I give up – I know you won't budge when you're set on something. But the staff also determined it likely that in a few centuries the people on Earth are going to send shiploads of young men and women to land on a narrow isthmus between two continents and build a canal that will link two oceans."

"Good for them. It's always good to keep organisms busy. I know it works for beetles."

"Yes, but these humans have never been exposed to Stegomyia mosquitoes whose bite will give them Yellow Fever and in short order thousands of these young people will die – sometimes in just days. They have no way of knowing the danger. We need to give them a warning of what is to come."

"Honor your mother and father."

"Excuse me?"

"A good Commandment for this situation."

"In all honesty Madame Creator, I don't see how that can prevent these imminent deaths."

"Didn't say it would and I'm not even sure what it means but it has a nice ring to it – one of my best – I like it a lot."

"Madame Creator, we're talking about the lives of thousands of young people."

"I'm sure – look, Ron, after the new beetle, I still want to create a galaxy or two this afternoon – so what else you got?"

"Well the staff also found that the bite of another mosquito, the Anopheles, may condemn millions of humans to malaria which means a lifetime of recurrent fever, shivering, vomiting, retinal damage and convulsions. We've got to get a warning out to them. By the way, why are you so fond of these mosquitoes – maybe you shouldn't have created them?"

"I don't know – I think they're kind of neat. Not up to beetles you understand, but still nifty."

"Wouldn't it have benefited humanity to have created more beetles and less mosquitoes?"

"Hmmm – never thought about that. Anyways, just remind humans of this one. "Don't bear false witness"."

"I'm sorry Madame Creator but this is a serious problem that will ruin the lives of millions of humans on this little planet. I don't think another catchy Commandment is going to alleviate their suffering. I'm not even sure what this one means."

"That's why I like it – it'll keep them guessing and meanwhile they don't cause trouble - they stay quiet. Anyway, I think you get too serious Ron, you need to learn to relax."

"Yes but back to malaria – we need to inform humanity and prepare them so they know what is coming and what they can do to prevent all this suffering."

"Look, they need to learn on their own how to react to problems. If I prepare them ahead of time, they'll never learn to take care of themselves. Besides, one little planet more or less isn't really going to affect my Universe. You know how many galaxies I've already created?"

"No but that is something else the staff discussed. We have billions of them already. Isn't that enough? We need to conserve our assets. Who needs all of these galaxies anyway?"

"It's not a matter of need – I just like them. Keeps me busy. I make all types – ellipticals, spirals, lenticulars, irregulars – really neat stuff."

"But there're already billions of them."

"Look, Ron, we all have our little habits – this one's mine. Is it hurting you? Anyway, I've got to run – what else you got?"

"As I said, my staff came up with catastrophes that may await the people on this little planet Earth. Here's another that especially caught my attention. Humans are prone to getting injured and thousands will die because their doctors are unaware of the need for cleanliness when performing the requisite surgeries. The staff predicts some of these humans will have a civil war during which they will perform battlefield surgeries with little regard for sanitation – lucky if the surgeons even wash their hands - you can imagine the suffering and deaths from infection caused by all the nasty little bacteria and viruses you created."

"Keep this up Ron and you'll ruin my lunch - I've never been able to stomach those little thingys!"

"Madame Creator!"

"Oh, all right" "No graven images"."

"What? What's graven mean?"

"Carved."

"What will this Commandment do to stop infections?"

"Probably nothing but they won't find that out for years and it'll keep them from complaining in the meantime."

"These maxims of yours remind me of some the staff came up with to improve life for humans on Earth. Would you like to hear some of them?"

"No, but I'll get out of here faster if you spit them out."

"Care for the poor, the less fortunate, and the ill. Feed the hungry. Shelter the young and the aged. Protect the rights and freedoms of all with just laws. Encourage education for everyone. Treat all life as humanely as possible. Seek knowledge of the Universe through science, engineering and medicine. You will note these maxims are positive in nature. The staff is of the opinion that life on Earth would benefit more from positive goals than from negative rules. In any case, these are just a few that they came up with – I could go on if you wish."

"No coveting of your neighbor's possessions."

"Beg pardon?"

"No coveting. Crave, envy, desire, lust, none of that sort of thing."

"With all due respect, Madame Creator, this is another example of negative maxims the staff tried to avoid. Anyway, what does that have to do with the poor, the less fortunate, the ill and the aged?"

"Nothing, but I'm off for lunch in the Fornax Cluster Ron, and if you don't get out of my way, you're history."

Chapter 17

IT MAY SEEM PRESUMPTUOUS TO VISUALIZE THE UNIVERSE from the Creator's point of view - to study her past efforts – to consider her objectives—to imagine her future plans – to look into her thoughts concerning humans – in a manner of speaking, to put ourselves in her shoes or whatever she may use for shoes. On the other hand, investigating this line of thought might be productive. So to begin, we might first look at all she has on her plate. There is, of course, her overall Universe to consider. A part of this Universe is our Observable Universe. We have no way at the present of knowing the comparable sizes of these two entities. We do know our Observable Universe has a radius of approximately 13.7 billion light years. And we know the Creator has scattered across this Universe considerably more than two hundred billion galaxies. Finally, we know she has formed many of these galaxies to each contain at least one hundred billion stars. In other words, the Creator has a lot on her plate.

In Chapter 23, we estimate there is or has been intelligent life on at least ten billion planets in the Observable Universe. To be super conservative, let's further reduce this estimate by a factor of one million. Then the Creator could still, if she chose, interface with intelligent life on ten thousand planets. What thoughts or plans might she have relative to this group of planets sprinkled across the vastness of the Observable Universe? Since we have no knowledge of how the

Creator thinks, no way to appreciate her outlook, no way to understand her feelings or lack thereof, and no way to take measure of her objectives and plans, we must consider that all options are open to her.

A first option would be to do nothing at all. That is, to ignore these intelligent life forms. To leave them to their own devices. Perhaps the option of communicating never even enters into her plans. Perhaps these intelligent life forms are incidental byproducts or unintended consequences of which she is unaware or for which she cares not. After all, we have no idea of what the Creator's plans are or if we figure into them. If, however, the Creator does wish to communicate it would seem she would convey to the intelligent life on all of the ten thousand planets. It would not seem logical to select some and ignore others. Of course our grasp of logic may have little in common with what is logical to the Creator.

Assuming, however, the Creator does wish to communicate to these life forms, she might then consider the frequency of her communications. Should she communicate often during the time span of the intelligent life on any given planet? Should she communicate on a regular basis? Should she communicate only once? The last option seems a curious choice. Maybe that is what a Creator would do but it doesn't seem logical – at least not from a human point of view. If she were to convey only once, however, how and why would she select the time? Early in the span of the species? For example, early in the life span of Homo sapiens on the planet Earth? But why not communicate again or even often?

If the Creator does indeed decide to communicate only once, she would then have to decide at what point should the communication be made in the history of the species. It would seem this point would not be before the development of language and at least a rudimentary form of written disclosure. On the other hand, it would seem the Creator would convey as early as possible – surely she would not wish to deny her attention to any more creatures than necessary. Once the Creator has decided when to communicate, she would probably next consider how to make that communication. Then she might decide whether to communicate to men only, to women only, or to everyone. And should she leave a tangible object or reveal something unknown as proof that the communications are indeed from the Creator?

Just for fun, imagine we simple humans could tiptoe into her presence and listen as she considers options and makes decisions in her mind. Of course, that assumes the Creator is structured with a mind. Just because we are, there is no reason to make such an assumption. Perhaps she has other decision-making structures. Perhaps she doesn't even have physical structures in the sense that we do. It also assumes the Creator considers various options in order to make decisions. Perhaps yes and perhaps no. Is it not probable she works in ways that differ from ours? These questions we cannot answer. We have no evidence one way or the other and we have no way of submitting questions nor do we have any avenues of research open to us.

But it would seem she must have some powerful decision-making avenues open to her. After all, she formed the mighty Universe with its forces of gravity, electromagnetism and the weak and strong forces and its laws of motion, thermodynamics, energy, and heat. Let us not philosophize further as these issues are obviously beyond us. Rather, let us simply listen in as best we can to her decision processes and apologize ahead of time for our intrusion into her affairs.

"OK, now that my desk is clear of Revelations to those really sharp level nine critters on that lovely little planet out in the Bootes supercluster, let's she what's next on my list. H'mmm, looking down the list here – oh, yes, a planet called Earth with level seven thingys known as humans. That's really below my normal cutoff limit but what the hay - I've got the rest of the decade free. I'll just wrap this one up and then call it an eon".

"First thing I always decide in these affairs is what set of Revelations to use and when to convey them. Hmmm, level seven, not too swift – well, no use wasting something fancy on them - the trusty old Commandments should hold them until they sharpen up in a few thousand revolutions of the Earth about its star – what do they call those? – let's see here, looks like its "years". Now, next decision, when should I make a visit?"

"Hmmm - my calendar is pretty full for the next hundred years or so but I could pencil them in after that. Of course, I could put it off for a few thousand years – I think I've earned some time off after that communication to the level nine critters out in the Bootes. On the other hand, I could have penciled in these human thingys a thousand years ago. Of course, those humans that lived before my Revelations don't benefit from them. So what's the best time?"

"Well, that's always a sticky question and there's really not a best answer. Earlier is better, of course, unless they're just not advanced enough for it. Later means they're more certain to be able to make good use of the Revelations but of course that means a lot of them are left out and what happens to them? I always wondered about that but haven't ever been able to research it fully. I don't like to think about it - it's not that important – let's not worry myself sick over it – I make too much of these decisions as it is. I'll just work them in after that scheduled visit to those level eight critters out in the Coma supercluster."

"OK, now I've worked that out, how shall I convey the Commandments? That's always a big decision – so many ways. Hmmm – I've been working pretty hard lately so maybe this time I'll just take the easy way out and communicate through one of their males in private and ask him to pass them on – then I'm out of there – that's the fastest, neatest way. I don't know though – I've done that before and the other critters on the planet tend not to believe the one I select. They say he could just be faking it - I can see their point. Not much better if I pick one woman. Better if I gather a group but why stop there? Hey, I am the Creator, right? – I've got ways to reach every single being on the planet simultaneously."

"I recall the old days – for example, that time out in the Shapley supercluster when I put up a huge billboard in the sky and just let the planet turn beneath it. That worked great – every being could read my Revelations once every rotation of the planet. Of course, everything's got a downside – they complained - got a bit tired of seeing it every day and said it cast a shadow - still that's their problem – I can't worry about everything in the Universe. And then that time in the Hydra supercluster when I appeared simultaneously in front of every being on the planet and gave my pitch in person – mano a mano, so to speak – whatever that means. In three dimensional color yet – boy, that was something but it took a lot out of me – my batteries needed recharging after that and it was a couple of eons before I really was back in the swing of things."

"Now, how do I insure they know it's me, the big Creator? You'd think that would be obvious but on the other hand, I understand – it could just be a story that gets passed around – how would they know the difference – there's always someone to start up a rumor. So, let's see – well, I could leave some tablets or plates printed with the Revelations but that means I have to research their languages and repeat the Revelations in at least a few of them."

"I could fashion them out of an element they're not familiar with – I've got a bunch of those at the end of what the humans call the Periodic Table of Elements and that would be convincing. Last time I did that though, the radiation killed a few critters – that wasn't too cool. I could create a new moon and inject it into orbit with the message on its face. That would be impressive but it might alter the Earth's axis and rotation rate and eject a few million humans into space – probably not a good thing if that got out. I could instead position a small black hole nearby – h'mmm, impressive maybe but what good would that do and it might suck in a few million humans – bad public relations. Well, I'll get back to that."

"Now, what's next on my list? Oh, yeah – should I tell them something they don't know? When they research it out later, there will be no doubt in anyone's mind it was me – the real deal. These humans are level seven so I don't have to get very advanced or go too far out. I could just mention that the Earth is round and revolves about the Sun. Oh, wait a minute – I see in the big Universe record book they know that. OK, I'll just mention that all elements consist of atoms and electrons and that light has a velocity independent of the speed of the observer. What the hey! – they know that too! How about the genetic code? That's really complicated and always a big winner. Wow, it says here they figured that one out too! The Universe records must be mixed up – these human thingys are way past level seven. OK, after lunch, I'll have to do some more work on that one"

"What's left to do? Oh, yes, how often should I communicate to them? I've tried various options on this one. Every billion years out in the Hercules supercluster with those strange round critters with legs sticking out at all angles. I've only gotten back two or three times to the level nine beings in the Bootes supercluster – there're so advanced they like to correct me and that makes me a little uneasy and a bit cranky. And the level eight critters in the Leo supercluster with the two heads that look forward and backward – that's fascinating so I'm sort of fond of them and drop in every eon or so."

"I'll have to give that some more thought and --- Holy Universe– look at the time! I got so much into these human critters on Earth and how to approach them I completely lost track of everything else. I can finish up on this later – right now, I'm out of here – I haven't missed out on that happy hour out in the Sculptor supercluster for eons. I love the dip and the free glasses of liquid hydrogen. Later, humans – by the way, you didn't fool me – I knew you were listening in – but no harm, no foul – take care – it's a dangerous Universe!"

Chapter 18

IMAGINE SCENARIOS A AND B CONCERNING THE COMMUNICATION OF A CREATOR'S REVELATIONS. In scenario A, the Creator voyages across the vast spaces of her Universe and privately communicates her Revelations through just one man of all the people then living on Earth. The Creator, however, fails to leave a tangible object as proof of her visit and fails to disclose anything not known at the time of her communication.

Although this communication is only between the Creator and the selected man, the Creator commands, either implicitly or explicitly, that he should pass these Revelations on to others. The "others" may be interpreted to be other members of the man's tribe or society or may be interpreted to be all of humanity. In either case, over time the Revelations become part of the religious teachings of a substantial portion of the peoples of Earth. In scenario A, the Revelations were indeed received from the Creator.

In scenario B, various men of a male-dominated tribe or society begin to originate stories concerning the existence of a Creator and concerning Revelations from that Creator. This story-based process may help them deal with the stresses and sorrows of life's journeys. Belief in a caring Creator can help to ease the cares and pains of life such as hunger, illness, regret and loneliness.

In any case, succeeding generations then modify, embellish,

and add to these stories. Over time, this process leads to a final written version which is handed down generation by generation. At some point, the details of the origination process become lost in history and the Revelations become part of the religious teachings of a substantial portion of the peoples of Earth. The Revelations were indeed invented by people but the process of invention was over many generations and a long course of time so that the origins of the process were lost.

Upon reflection, it may become apparent that later generations cannot distinguish between scenarios A and B and thus have no way of determining which occurred. Imagine the Creator had, instead, met with a group of men and women and imagine that each of the group had written an account of the meeting and imagine the accounts agreed on most substantive details of the meeting and finally imagine numerous ones of the accounts have survived to the present day.

They would be most persuasive because they came from several people and they agreed in most details. This would be a strong argument against collusion and this would tend to convince a large portion of humanity that the Creator had indeed visited Earth and that the surviving Revelations were the words of the Creator. The accounts would be a strong indication of scenario A.

But since we were not there at the time and it is no longer possible to interview the Creator nor those who passed on his story, we have no way to verify this account. One man may be honest, may be sincere but deluded, or may be dishonest so that testimony from one man is not convincing. In the case of the Revelations we lack testimony from others since the communications were restricted to be through one man. Accordingly, there is no way to distinguish between scenarios A and B. Either may have occurred but, because the Creator only communicated through one man, it is impossible, on this basis alone, to determine which of scenarios A and B truly happened.

In contrast, imagine the Creator had left a tangible object as proof of her disclosure to one man and that this object could only have come from the Creator. Judaism teaches that the Creator gave Moses stone tablets and Mormonism teaches that the angel Gabriel gave gold plates to Joseph Smith. However, none of these tangible objects can now be produced. It is said the stone tablets disappeared when the second temple was destroyed around 70 CE and it is said the gold plates were returned to the angel Moroni around 1829.

But imagine the Creator left another tangible object. Suppose, for example, it was made of an element that was not one of the ninety two naturally-occurring elements on Earth. As humanity became more scientifically advanced, it would have been more and more convincing that this object had come from the Creator. Because it would have been the only example of this element on Earth, most people would have been thoroughly convinced of the Creator's visit. However, humanity is not aware of such an object nor does any known religion teach that the Creator left such an object.

Figure 17
"later generations cannot distinguish between scenarios A and B"

Other than the stone tablets and the gold plates which cannot now be produced, it is not among the teachings of any of the major religions that the Creator left an object whose origin was not an Earthly one. There is, then, no way to distinguish between scenarios A and B by this account. Either may have occurred. Because the Creator never left on Earth a tangible object having an origin other than the Earth, it is impossible on this basis alone, to determine which of scenarios A and B occurred.

Imagine yet again that the Creator had disclosed something unknown to humanity at the time of her Revelations. For example, suppose the Creator told Moses that the Earth was round or suppose the Creator disclosed through Jesus that the Earth revolved about the sun. Suppose the angel Gabriel told Muhammad that the planets moved about the Sun in elliptical orbits or suppose the angel Moroni told Joseph Smith of the presence of the outer planet Neptune. In each of these example, the information was not known to anyone on Earth at the time and its disclosure would have assured humanity in ages to come that the Creator had visited the Earth.

But the Creator did not do so. There is no record of the Creator ever having disclosed to humanity anything not known to humans at the time of the disclosure. One might have thought the Creator would have disclosed some knowledge of this type almost by accident. Since she knew so many things about nature, the Earth, the Sun, the Milky Way, and the rest of the Universe, it might have seemed she would have inadvertently mentioned something of which humans were unaware. On the other hand, being all-wise and knowing that humans were naturally skeptical, it would also seem the Creator would intentionally mention something unknown to assure that humanity would know she was the source of the Revelations.

Had the Creator disclosed the principles of a telescope to Moses, had she told Jesus that Moons orbited the planet Jupiter, had the angel Gabriel informed Muhammad that the Milky Way was a gigantic disk of stars and that the Earth was one of them, or had the angel Moroni told Joseph Smith that there are planets about other stars of the Universe, humanity would have known later that they had heard from the Creator. None of these events, however, occurred.

Because the Creator communicated to only one man, never left a tangible object, and never disclosed anything not known at the time of her communication, there is no way to distinguish between scenarios A and B. Either may have occurred. On the other hand, either may be false. It may be that the Creator truly communicated Revelations to humanity. It may also be the that the disclosures did not come from the Creator but simply grew over time as initial generations originated elements of the story and succeeding generations added to it and modified it until it assumed its present form. The fog of time then grew dense enough to hide the origination process.

Because we cannot interview members of these generations, we have no way to determine if the present form of the Revelations developed through scenario A or scenario B. Some people may claim that scenario A is the true one but there is no way they can prove that claim. They can only urge that it be accepted on faith. Similarly, those who claim scenario B is the true one have no way of proving that claim. They can only say it is more likely to be the truth.

Occam's razor is sometimes used to indicate which of different theories is most likely to be true. This is a principle of reasoning attributed to the 14th century logician and Franciscan friar William of Ockham. One version of this principle is "given two theories that make the same predictions, the simpler is the better" and another is "one should not increase, beyond what is necessary, the number of entities required to explain anything". Occam's razor is sometimes called the principle of parsimony because it urges one to choose the simplest from otherwise similar theories.

Occam's razor could be used to examine, for example, the crop circles of matted grass or crops that began appearing in the 1970's in England and various other countries. One theory was that unidentified flying objects (UFOs) were responsible. Another theory was that humans pushed down the grass or crop with various tools. It is unlikely a UFO could travel at the speed of light and it is unlikely that intelligent life is on the nearest star but if these unlikely possibilities were true and this life produced and launched a UFO towards Earth, it would have to make a journey of over four years to reach Earth.

Forgetting other difficulties of such a journey, simply providing the UFO with enough food would be a daunting task. In contrast, it would be relatively simple for humans to make the circles.

As to motive to make the crop circles, it would seem that extraterrestrials would have none. They might be motivated to travel through space, even travel to our Earth, but once here, there seems little reason why they would want to make crop circles. What form of life would make such a stupendously difficult journey to tramp down crops? On the other hand, humans have a long history of performing pranks for various reasons which include just the love of doing something outrageous.

In this case, the application of Occam's razor would select the simpler theory of human responsibility as being the more likely theory to be true. And, in fact, two pranksters admitted in 1991 they had made the original crop circles with simple tools. In a demonstration they made a crop circle in one hour. Others have since made a variety of complex patterns using tools such as ropes, wooden planks, plastic pipes and ladders.

Scenario A proposes that the Creator disclosed Revelations to one man but never left a tangible object and never disclosed anything not known at the time of her disclosure. Scenario B proposes that humans invented tales in which the Creator disclosed Revelations. Some might use Occam's razor to select scenario B because scenario A requires the presence of a caring Creator and that is an unlikely event given that communications from her have never been confirmed. Others might use Occam's razor to select scenario A because scenario B requires the invention by humans of some rather complex stories and requires that this process was never exposed.

In fact, there is no way to prove which of scenarios A and B occurred. Either may have been the source of the present written and oral religious teachings but we have no way to definitively disqualify one or to definitively prove the other. Those who believe scenario A say those who believe scenario B are heathens and those who believe scenario B say those who believe scenario A are misguided and naive.

Chapter 19

ALTHOUGH SHE MAY HAVE FORMED THE UNIVERSE, it does not necessarily follow that the Creator can easily visit various portions of it. At its present age of 13.7 billion years, the Universe is so vast it may not be possible for even the Creator to simply drop in on any supercluster she might choose. The enormous distances became apparent in Chapter 3 where we mentally voyaged about the cosmic web. As noted there, Earth's closest visible star is Alpha Centauri A which orbits about a smaller star Alpha Centauri B with a third dwarf star completing a three-star system. Alpha Centauri A is 4.3 light years from Earth. Some feel for this distance may follow from noting that if the span between the Earth and the Sun were scaled down to one meter, then Alpha Centauri A would be 169 miles away

The Voyager 1 spacecraft was launched from Earth on September 1, 1977 to study the outer solar system. Its path was selected to include a gravity assist maneuver which would enhance the spacecraft's speed. Also known as a slingshot maneuver, this technique directs a spacecraft past the edge of a large planet. The spacecraft is accelerated by the gravitational pull of the planet and its path is redirected as it leaves the region of the planet. This maneuver changes the spaceship's trajectory and speed relative to the Sun and, because the spacecraft appears to have bounced off the planet, it is sometimes called an elastic collision.

After Voyager 1 flew by Jupiter on March 5, 1979, it used the gravity assist technique as it passed by Saturn on Nov. 12, 1980. Accordingly, its path was bent about Saturn and directed away from the solar system's ecliptic plane which is the plane in which the planets orbit the Sun. In early 1998, Voyager 1 departed the Solar System at a speed approximating 39,000 miles per hour. At that speed and if Voyager's path was properly oriented, it could reach Alpha Centauri A in approximately 72,000 years. For this journey, even the Creator might want to pack a light lunch or two and an extra pair of pajamas.

The fastest mode of space travel that is presently envisioned uses the controlled power of pulsed nuclear explosions. If this propulsion technique can be perfected in the future, it is anticipated that it might achieve speeds on the order of 5% of the speed of light. This would reduce the travel time to Alpha Centauri A to mere 85 years. The Creator would still need a lot of good books to read and a crossword puzzle or two.

These examples of space travel emphasize the daunting time durations required to voyage about our Observable Universe. We previously followed a journey of the Creator that brought her to our Virgo supercluster after which a trek of thirty or forty million light years found her at our Local Group of galaxies. A final inward jaunt of a million or so light years carried her to our Milky Way galaxy. Then a brisk sprint of a few thousand light years along the Orion spur brought her to Earth. It seems most unlikely the Creator ignores the rest of the mighty Universe and restricts her attentions to our Milky Way galaxy alone. But even if that were the case, it would take her 100,000 years to simply stroll, at the speed of light, from one side of our galaxy to the other.

Then is a visit to Earth by the Creator impossible? The journeys above were described in accordance with our known laws and principles of science because this provides a feel for the incredible size and structure of our Observable Universe. However, this is the Creator after all. Surely she has some more scientific rules up her sleeve than those we have been able to discern after several thousand years of endeavor. She probably is not limited to our known forces of gravity, electromagnetism and the weak and strong forces and our known laws of motion, thermodynamics, energy, and heat.

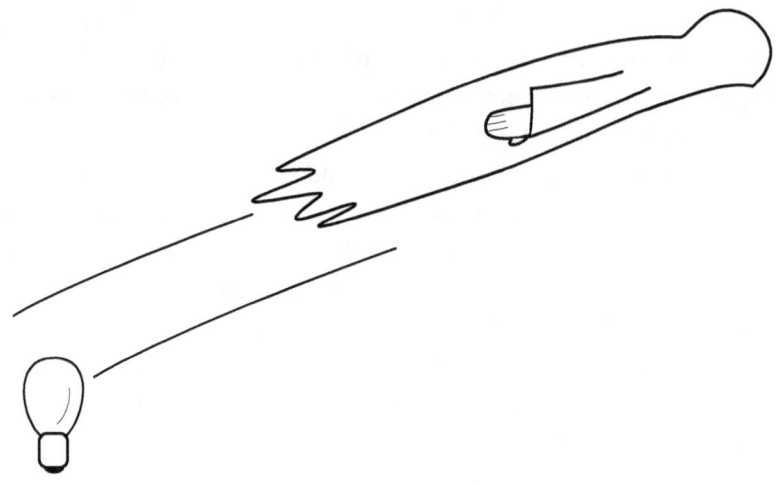

Figure 18
"perhaps the Creator is not limited to the speed of light"

The fastest known mode of communication between different portions of the Observable Universe is the speed of light. It may well be, however, that the Creator is not limited to travel methods of which we are aware and that she can cross the vast spaces of the Observable Universe in times far less than those described above. Perhaps, then, the Creator is not limited to the speed of light. Perhaps she can even pass instantly from one Universe location to another so that she can appear at one and then simply appear the next moment at the other without physically transitioning the space between. Perhaps she can even appear at different sites simultaneously. We have no way of knowing the capabilities and the limitations of the Creator so it is entirely possible she can travel faster than the speed of light.

We do know that our knowledge of the Universe was primitive as late as the times of Muhammad but that it has grown enormously since the era of Joseph Smith. Humans living just before Joseph Smith's death in 1844 would have known, for example, that each chemical element is composed of a unique type of atom, and that there is some relationship between magnetism and light. They would not have known, however, that electricity, magnetism and light are all manifestations of electromagnetism, that electric and magnetic fields travel through space in the form of waves at the constant speed of light, and that the speed of light is the same for all observers regardless of the state of motion of the source. Nor would they have known of electrons, quantum physics, that mass and energy are equivalent concepts, and that, in each element, the nucleus of the atom is circled by a respective number of electrons.

Humans living before 1844 would have known a bit about the shape of the Milky Way galaxy but very little of the position of our Sun in the galaxy. And they would not have known that there are other galaxies with spiral, barred spiral, elliptical, and irregular shapes, that galaxies are grouped in clusters, and that there are super clusters which contain thousands of galaxies. And they would not have known that there are black holes at the centers of many galaxies, that the size of the Universe is expanding, nor that the radius of the Observable Universe is on the order of 13.7 billion light years.

Our knowledge of the Universe will continue to grow at an accelerating rate. Many concepts will eventually become apparent that we can not anticipate from our point in time. These will undoubtedly alter our present understanding and comprehension of the Universe. Perhaps we will even learn something of the structures outside our Observable Universe. We may assume they are similar to those of our portion of the Universe but nothing is assured and we have found in the past that our assumptions were not always correct. Then is it possible we will even learn the Creator is not limited to the speed of light in her travels across the Universe? Stay tuned.

Chapter 20

"**WOW!** I'VE SEEN SOME IMPRESSIVE OFFICES but nothing like this! All this gorgeous antique furniture.! And the view – magnificent! They don't call you the Creator for nothing."

"OK, enough pleasantries, Penny – let's get down to work. I guess you know why I called you in?"

"Well, I got the impression you wanted to update an act you played some centuries back. Not many actors have a gig that spans centuries – pretty impressive."

"Look, I'm not an actor, I'm the Creator – you do know the difference?

"Got it – so what's up?"

"I thought my original Commandments were for the Ages but times have changed. I come back from a loop around the galaxies - I hate traveling by the way, I'm always losing my luggage - to find that these people on that little planet Earth are completely different from the ones I dealt with before. Back in the day, people were unaware of others a couple of hundred miles away and they didn't even know the Earth was round. And they didn't ask questions – put out a Commandment or two and then just sit back and enjoy. Almost as much fun as creating a new species of beetle. You got respect if you know what I mean. Now I come back from my tours across the Universe and find it's all instant communications here – they're all tied in together on an internet thingy – something happens in one place and it's known everywhere the next moment. You can't hide anything – everything's on talk shows – and there's no respect for authority. I come back to find people are even expressing some doubts about my original Commandments – that maybe they didn't come from me – that maybe I don't even exist. It's like in that Wizard of Oz thingy where Toto pulled back the curtain – these people want proof for everything."

"What's this Oz thingy?"

"Something I picked up on from Chad."

"Oh, OK. He's full of those associations. Anyway, I've heard some doubts myself. So what do you want from me?"

"I've decided to do an update on the Commandments bit but I know we face a different audience, a different scene and a different time – we can't just do the same ol' same ol' – we need new ideas that will work in this new world of communications. I know I need a different approach and I hear you're good at presentations!"

"I'm the best if I do say so! My work is all over the internet."

"What's this internet thingy?"

"It's kind of like a giant electronic bulletin board."

"Electrons I know, Penny – need I remind you I invented them. But what's a bulletin board."

"You don't get out much, do you Chief – sorry, Creator?"

"Hey, give me a break – I've been out in the Columba supercluster for a long time converting heathens."

"Sorry, I know you have a whole Universe to look after."

"OK, skip it. Look, I'll come up with the Commandments – that's my thing. It's not that they're so different from before as much as it is I know we need to push them differently in this new world and that's where you come in. I've seen some of the work you do."

"Did you have any thoughts?"

"Well, I figure we'll drop one of the glossiest brochures in all of history on one man – maybe send it over that internet thingy of yours. I have sources – I can fund a virtually-unlimited budget."

"Chad said you had budget problems."

"I've found some new funding."

"OK, but I don't think that will work, dude."

"What's this dude thing, Penny?"

"Oh, sorry! I'm thinking ahead already and I forgot you like to be referred to as the Creator."

"That's probably because I am the Creator."

"Right – look, I don't think one glossy brochure to one dude is going to swing it and if there's the slightest error it's over the internet and then it's everywhere the next morning. Anyway, people here on Earth have been through all this over the ages – they're pretty knowledgeable – a bit jaded - they sense when they're being hustled."

"What's this hustled?"

"You know, picked up, procured, that sort of thing."

"I'm not following you."

"Forget it – I'm thinking here – OK, how about this? We get you on all the big shows – interviews – that sort of thing. Then we go on tour – Vegas, London, Hong Kong, smoke and mirrors, the whole ball of wax."

"I'm the Creator – I don't do interviews – I hand down Commandments."

"You don't do interviews? Why not?"

"How would it look – some dame or dude asking the Creator what she has for breakfast? Does she workout? "You look buff - which diet do you follow?""

"How come you know buff and you don't know hustled?"

"Let's drop it, Penny! No interviews!"

"This is going to be tougher than I thought - give me some space here – OK, let's try this - we'll buy an hour's time on every major world network and we'll have it streaming over the Internet – ahead of time, we'll advertise for months that the Creator is coming out with updated Commandments. On the scheduled date, you'll descend from the heavens with clouds and light rays and rolls of thunder in the background – the biggest stage of all time and the biggest audience of all time. No questions, no interviews, no moderator, no introduction. Just you booming out your new Commandments – I know some great sound men – they'll do some voice reverbs and we'll have the occasional lightning flash in the background for accent."

"What's reverbs?"

"You know – echoes – that sort of thing – it adds substance and dignity – no offense but I've heard your voice – a bit thin."

"I don't know, this sounds like something Chad came up with – I've always liked just one on one – you know, just me and a nice, tall good looking fellow with a beard – then let him pass on the Commandments just like in the old days."

"What's this hang up on beards and I thought you wanted to update your act?"

"The act is fine – I just don't want to lose my, what do you call them, ratings?"

"You want ratings? How about this? We have a huge stage in front of you and heavy curtains behind. Maybe flags of all the nations arranged in front of the curtains. Anyway, we have a long lineup of humans – young, old, tall, short, kids, adults, well-dressed, street people – the whole nine yards. One at a time they come out in front of you and you give each of them a copy of the Commandments straight to the shoulder. One after the other for a straight hour with only two breaks for ads and the rest room."

"The ultimate in tension and human interest and all the time a worldwide audience is hearing your Commandments read over and over and over – it should get the highest television and internet ratings of all time and no one watching will ever forget it. There will be no doubts – every viewer will be convinced this is the real thing – they may even etch your thoughts in stone and put them in every state house in the world. There would be nothing between your Commandments and every viewer across the world – instant one-on-one communication - the next day, every word of your Commandments on every tongue. Where are you - you've been pretty quiet – what do you think?"

"What? Sorry, did I miss something, Penny? I was visualizing myself and one man going mano-a-mano on that internet thingy of yours. Probably look best if I wore one of the big robes – like up there on that ceiling thingy, you know?"

"You didn't hear a thing, did you?"

"No, no – you're doing great work – it's just that I remember the old days – I got a lot of respect back then. Unbelievable, how soon they forget."

"Let me make one last try to spell this out, Chief. Once they've been tied into a worldwide internet with instant news and communications, this new bunch isn't going to buy the old story of one man receiving Commandments from his Creator up on mountain tops, on stone tablets, on buried gold plates, whatever – they want to be included – they want to be connected - they want to be part of the story. And the women don't stand behind the men like in days of yore. Now they're pushy – up front - they want theirs."

"I'm all ears for the little dears. Anyways, that's why I want to bring out a new set of Commandments."

"OK, I know women – after all, I am a woman - here's what we do to keep them happy. You sit down in one of their afternoon television shows with a few ladies arranged in front of a small audience of more ladies. Play it light – a few jokes - a little banter – show your human side, so to speak. Then, at the right moment, slip in your new Commandments so that the entire audience receives them straight from the Creator."

"That's good, Penny. I could even lighten up a bit – get down on their level and congratulate them on working out and staying buff while also maintaining a home, helping the kids in their school work, keeping the husband happy, and taking the dog out for a walk – show them I understand how busy they are in this new world of – how did you put it? – oh, yeah, instant news and communications."

"By the way – ditch the big robe – better if you're up-to-date – maybe just wear a smart pantsuit."

"No pantsuit, Penny – I have to draw the line somewhere,

maintain some decorum, and keep some distance from these human critters."

"OK, but I like your idea of relating to their different tasks - that will show them you know all about multitasking."

"Is that like when I create more beetles?"

"No, no – that's their word for doing several things all at the same time."

"Oh, like in the past when I created dinosaurs, worked out electromagnetics, moved continents around, invented black holes, organized galaxies and came up with supernovas."

Figure 19
"sit down in one of their afternoon television shows with a few ladies"

"Yes - by the way, how did you come up with the concepts of supernovas?"

"Never mind. Tell you what – I'll go through with it."

"Great!"

"Just a few changes."

"Oh boy!"

"I have no trouble with the banter, Penny – I can banter with

the best of them. As part of it, I'll kid them along – tell them my
Commandments are more of a man thingy – that woman's place is in
the kitchen – that women should be barefoot and pregnant. Then when
they see I can banter, that's the time to get serious. Tell them to look on
that internet thingy to see me in the big robe. Wait – this just doesn't
feel right – I don't know - it's not me - I liked it better in the old days
before this world was all connected. Thanks for your thoughts Penny,
and send me your bill but I have to do it my way – like up on a
mountain top with just one man – you know what I mean?"

"I'm out of here!"

Chapter 21

IF THE CREATOR HAS A SPECIAL INTEREST IN HUMANS, is it not curious she would begin forming our solar system from the gas and dust of a molecular cloud some 4.5 billion years ago but not get around to developing multicellular life forms on Earth for about 2 billion years and would then wait until just 2.3 million years ago to introduce the genus Homo? Would she then delay until almost 200,000 years ago before evolving this genus into our species Homo sapiens? In fact, if we were to represent the 4.5 billion year history of our solar system as a 24 hour day as in Figure 19, then Homo sapiens would have existed during a bit less than the last 4 seconds of that day. It would seem the Creator either has a lot of patience when working on her plans for Homo sapiens or Homo sapiens ranked awfully low on her "to do" list.

Consider the Hadean eon (i.e., the eon from Hell) - the first eon in the development of the Earth. It began approximately 4.5 billion years ago (Ga) when elements of a giant cloud of gas came together to form the Sun and the planets. The Earth was then under relentless bombardment by comets, asteroids and other matter. Early in the Hadean, a collision with another growing stellar body may have tilted the Earth's axis and cast a great deal of matter into orbit where it eventually coalesced to form the Moon. In any event, the Moon was only 10,00 miles distant from the Earth in the Hadean. It is presently

239,000 miles distant and is moving away from the Earth by about 1.5 inches a year.

In this eon, the Sun was 70% as bright as it is now and the Earth was spinning so fast that each day was was only about 14 hours in duration. Molten iron sank to form the Earth's core with lighter elements rising to form the rocks of the outer surface. The oldest surviving ones of these rocks date back to a time just prior to the end of the Hadean. During most of this eon, Earth's atmosphere was primarily comprised of methane, ammonia, carbon dioxide and water vapor. Oxygen was either absent or present only in small quantities.

The rate of incoming matter seems to have intensified late in the eon during a period termed the Late Heavy Bombardment. This process may have been initiated by debris rained down from the Moon after it was struck by another celestial object. This bombardment probably kept the surface of the Earth in a semi-molten state and, accordingly, the name of this eon is taken from the word Hades which is a Greek translation of the Hebrew word for hell.

The Hadean eon lasted for 700 million years. Not only were humans absent during this great span of time but all life was absent. Obviously, the Creator had no plans for humans during this eon and was content to cool her heels along with the planet at this time. Because she was the Creator, one assumes she could have arranged the Earth during this eon to be a suitable home for humans but did not choose to do so. Apparently she had her reasons.

The Hadean eon ended 3.8 billion years ago. During the subsequent Archean eon (i.e., ancient eon) that lasted 1.3 billion years, Earth's continental plates began to form, there was intense volcanic activity and a magnetic field was established. The length of the day increased to 15 hours and the Sun's brightness increased to 80% of its current value. Many of the original components of the atmosphere escaped or were altered so that the atmosphere was now primarily nitrogen and carbon dioxide. The continents began to take shape and evidence of running water is preserved in rocks in Greenland that have been dated to 3.8 billion years ago.

Beginning around this same time, ancestors of today's blue-green algae began the process of photosynthesis which reconverted light energy to chemical energy. Although life may have developed as early as 3.5 billion years ago, definite evidence of microfossils has been dated to 2.7 billion years ago. At this time, most organisms were simple prokaryotes which generally consist of a single cell that contains a simple loop of DNA. Several ribosomes are provided to synthesize proteins with the aid of information stored in the DNA loop.

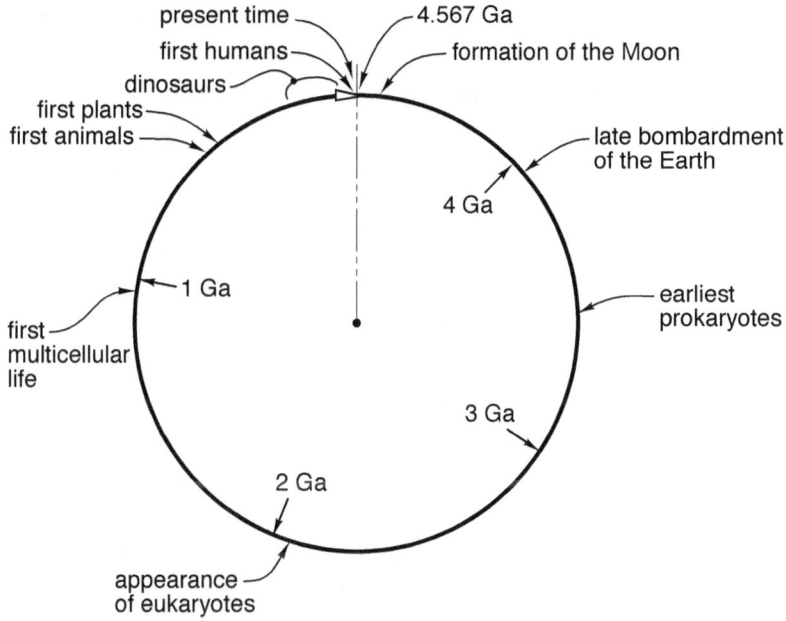

Figure 20
"represent the 4.5 billion year history of our solar system as a 24 hour day"

During the 1300 million years of the Archean eon, humans were still conspicuous by their absence. Of course they could not have lived in the Archean atmosphere but the Creator could have altered this atmosphere had she been so disposed. One can only surmise that the

Creator was busy elsewhere and other tasks were more important to her than forming a suitable environment for humans.

It was once thought that life on Earth originated around 2.5. billion years ago and so the beginning of the next eon was set at that date and given the name Proterozoic which comes from Greek words meaning "first life or early life". It is now known that life was already present prior to the beginning of the Proterozoic. Originally, this life consisted of the simple prokaryotic cells but more complex eukaryote cells appeared around 2.1 billion years ago.

Eukaryotes have a far more complex DNA that is housed within a nucleus formed with a double membrane. In addition to the nucleus, a variety of other organelles carry on the functions of the cell. For example, chloroplasts synthesize sugars, ribosomes synthesize proteins, and mitochondria are configured to use metabolism to provide energy. The endoplasmic reticulum is a membrane system for transport of cell products within the cell and Golgi bodies package cell products into forms suitable for transport away from the cell. In other words, eukaryote cells are complex life-supporting factories.

During the Proterozoic eon oxygen content of the atmosphere increased above 10% and an ozone layer began to block the ultraviolet radiation from the Sun. The duration of a day increased to 20 hours and the distance of the Moon increased to over 220,000 miles. Multicellular organisms made an appearance around a billion years ago. Still no humans - had the Creator no plans for them or were her other duties too onerous to permit their development? One would have thought she could have brought them along by now but it was not to be now nor for another billion years.

Multicellular life suddenly became abundant 542 million years ago at the beginning of the Phanerozoic eon. Animals with a variety of different body plans surfaced. Insects, amphibians, plants, fishes and reptiles first appeared in subsequent periods. Ferns and seed-bearing plants appeared about 400 million years ago. Although mammals appeared as early as 200 million years ago, they were dominated early on by the dinosaurs which ruled the Earth for over 150 million years.

Spiked and plated herbivore dinosaurs appeared during the Jurassic period that began 213 million years ago. During this period, mammoth sauropods grew to be over 100 feet long and weigh over 100 tons. The fearsome Tyrannosaurs lived during the Cretaceous period that ended 65 million years ago. After the demise of the dinosaurs, mammals rapidly diversified and adapted to conditions on land and sea around the Earth. The Moon moved out to 230,000 miles from Earth and the length of the day increased to 22 hours.

Then early members of the Homo genus first walked out of Africa around 1.7 million years ago. This was Homo erectus, a species that lasted over 1.5 million years and eventually inhabited Indonesia, Java, Vietnam, China and India. Neanderthals subsequently made their appearance around 370,000 years ago and moved into Europe around 230,000 years ago. Still the Creator had not seen fit to bring us, Homo sapiens, onto the scene. What could she be waiting for? One might wonder what was her plan for us - assuming she had a plan? Was she going to make us wait forever? Let's get on with it!

Finally, after more than four billion years of Earth history, our species Homo sapiens appeared on Earth about 195,000 years ago. This appearance makes sense if it was the natural result of evolutionary processes. In this case, it seems the Creator is quite content to let life evolve as it will under the influence of natural selection. But it makes no sense at all if it is thought to result from the Creator's fondness for our species. If she had special plans for us would she not have formed the Earth appropriately for our comfort at the beginning of Earth's history? It would seem from the evidence that she wasn't at all anxious to bring us on to center stage. But why make us wait around for over four billion years? However, this is the Creator and she can do anything she wishes. Then again, maybe not.

Chapter 22

ONCE UPON A TIME, WE THOUGHT OUR LITTLE EARTH WAS FIXED IN SPACE and the sun and the moon rose in the east, traveled west across the sky during the day and then somehow magically reset back to the east to rise again next morning. There was a moon and a backdrop of lovely stars which were unrelated to the Sun. Then a great shock - we discovered our Sun was just another of these stars and, even worse, we were the ones doing the revolving. Then we found our Sun and the visible stars were all in a minor arm of a huge galaxy. And now we know there are billions of galaxies strewn across the Observable Universe. The size and importance of our existence on our little Earth has steadily gone down in direct proportion to our increased knowledge of the Universe.

None of this vast assembly of galaxies was known in the early part of the 20th century when astronomers were trying to determine Earth's position in the Milky Way. One of these astronomers was Harlow Shapley who dropped out of school when he completed the fifth grade. After working as a newspaper reporter he returned to school and eventually entered the University of Missouri in 1907 at the age of 22 with plans to study in their new school of journalism. When he found that school's opening had been postponed, it is said, in a possibly apocryphal tale, that he decided to study the first subject in an alphabetical list. But when he found he couldn't pronounce archeology, he picked the next subject which was astronomy.

After earning a doctorate at Princeton University, he began investigating globular clusters of stars from the Mount Wilson Observatory in California. He determined that these clusters were centered about a point 50,000 light years from Earth. If they were part of the Milky Way galaxy, then our galaxy was much larger than previously thought. And if they were arranged about the center of the galaxy, then the Earth was not at the center as had often been assumed. In the "Great Debate" of 1920, Shapley took the position that the clusters were indeed part of the Milky Way galaxy, that they defined the center of the galaxy and that the Earth was far from this center. In opposition, fellow astronomer Heber Curtis argued that our galaxy was much smaller and that the Earth was near the galaxy's center.

Faint spiral nebulae had also been observed and astronomers had long been trying to determine the nature of these structures which appeared as blurred spots of light in the optical telescopes of the day. Shapley took the position that the nebulae were simply clouds of gas and were also a part of the Milky Way galaxy. Again Curtis disagreed and argued that they were separate galaxies that were spaced far from our own. If this were true, then our galaxy was but one of many and the universe was significantly larger than previously thought.

These disputes could be resolved if the distance to the clusters and the nebulae could be accurately measured. But how could this be accomplished? It was known that the brightness of some stars varied over a time period. They were called Cephied variables and as early as 1908 Henrietta Swan Leavitt began to study a relationship between the period and the actual brightness of these stars. She determined that the actual brightness could be obtained from the observed period and that the observed brightness could then be used to determine the distance to that Cepheid. The less the observed brightness, the greater the distance.

Using the Cephied variables as measuring sticks, astronomer Edwin Hubble was then able to show in 1924 that the distance to the spiral nebula M31 was far greater than the extent of the Milky Way galaxy. Therefore M31 was not in the Milky Way but was, in fact, a separate galaxy of its own. So Curtis had been correct about the existence of other galaxies. But further investigations showed that Shapley had been correct about the size of the Milky Way and the location of Earth within the galaxy. Shapley went on to identify 76,000 galaxies in the southern sky and became one of the first astronomers to realize that galaxies were often arranged in clusters.

We learned in Chapter 3 that there are probably at least 200 billion galaxies in the Observable Universe and the number of stars in these galaxies are virtually beyond counting. But life will only exist on planets of these stars that include the elements necessary to form carbon-based organic molecules. We now know the processes that generate these elements. In the early portion of the Universe, protons and neutrons began to condense out of the enormously hot primal materials. Then ionized hydrogen began to combine with free electrons and massive stars were formed with the resulting hydrogen. The fusion processes of these stars then progressed to the point where they formed higher elements such as silicon, nickel and iron.

At some point during the production of iron these stars collapse and then instantly explode. The resultant enormous temperatures and pressures generate the higher elements all the way up to uranium. These supernova explosions cast the elements out into the Universe where they are then used to form other stars and their circling planets. Carbon-based life could not have developed during the early years of the Universe because these processes would not have had time to provide the necessary elements. Accordingly, carbon-based life has existed only during a later portion of the Universe, e.g., perhaps the last 10 billion years.

Meanwhile, here on Earth we learned that 99% of our body's mass is comprised of just six elements - hydrogen, carbon, nitrogen, oxygen, calcium and phosphorus. Carbon, in particular, has a half-filled outer shell so that there are four unpaired electrons. Other elements can, therefore, readily attach to these electrons to form a variety of carbon-based molecules which are characterized by the presence of carbon-hydrogen bonds and are generally organized around rings and chains of carbon atoms. When carbon is oxidized during respiratory processes it forms carbon dioxide which is a gas that is easily expelled from the lungs of Earth-based life forms.

The element silicon lies directly below carbon in the periodic table of elements so that it also has a half-filled outer shell which provides four unpaired electrons. In contrast to carbon, however, when silicon is exposed to oxygen it forms silicon dioxide which is organized in a lattice that would be difficult for a life form to dispose of through any conceivable form of respiratory structure. From this knowledge, it can then be reasonably assumed that all life forms in the Observable Universe are based on carbon-based organic molecules just as they are here on Earth.

And now we wonder if we are the only advanced life in the Universe or are there others? Because our Milky Way galaxy is 100,000 light years in width, it is apparent we can never search through it or even a substantial portion of it for signals of life. It would seem that at best we can only search through a tiny portion of the Orion Arm which branches away from the Sagittarius Arm of our galaxy and angles outward to terminate in the Perseus Arm. Our little planet is found roughly half way along this minor arm - some 26,000 light years from the center of the galaxy.

But we probably can't even search for life through all of the Orion Arm because it would take a light beam 3,500 years to travel across it and over 10,000 light years to travel along it. The furtherest we might possibly look out from Earth would seem to be something on the order of 250 light years. We would receive a reply from an advanced civilization at that distance in 500 years. Many generations would have passed here on Earth and the original message would have been a faint memory before we received confirmation we are not alone. It would seem difficult to search among stars that far away and basically impossible among stars even further out.

But many of the stars and planets within this distance will not be suitable for the development of life. Some stars are so huge they only last a billion years or, in extreme cases, less than 10 million years and these time spans are probably not sufficient to develop complex life. In addition, many of these stars are part of binary star systems in which orbit zones that are suitable for life are more limited than is the case with lone stars. And some stars will have only gaseous planets that are not suitable for life.

Many planets are so close to their star that the available water is burnt off. Others are so far from their star they are permanently frozen. And some planets are so small their gravity is not sufficient to hold onto to an atmosphere. Life can only develop on a planet whose rotation rate is sufficiently rapid that its water supply is not locked into frozen conditions for extensive time periods. It may also help to have a moon and a tilted axis so that there are seasons, summer, winter, spring and fall. This would increase the radiation that reaches the surface and increase the temperature variability. It may also be helpful to have a large outer planet similar to Jupiter to shield inner planets from devastating asteroid impacts.

At this point the existence of advanced life among the stars and our chances of communicating with that life must remain purely conjectural. But let's attempt to explore the subject to thereby gain an appreciation for the possibilities and probabilities. First, note that there are approximately 260,000 stars within 250 light years of Earth. Let us assume that only 10% have a circling planet with conditions, e.g., temperature, atmosphere and tilted axis, that would support the evolution of carbon-based life forms. Now, of these 26,000 stars how often do variation and selection processes evolve an intelligent life form? Such life evolved only in approximately the last 200,000 years or so on Earth and was absent for billions of years before that. It would seem that intelligent life is rare so that one in a hundred might be an rational, if rather optimistic, guesstimate. This would narrow the number of candidate stars down to 260.

So we have estimated that a number of advanced societies have probably developed nearby in the Orion Arm. But what of the temporal relationship between them and us? After all, we can never communicate with a society that once existed but does no longer. For example, we can never interface with a society that lived for a time five billion years ago but no longer exists. Nor can we communicate at this time with a society that will only exist in the future. To make some judgement here, we must assume an expected time span of advanced societies.

Unfortunately we have no past experiences to rely on. Perhaps some judgement of this, however, can be gained from the time span of mammals during the Cenozoic era here on Earth. This is the present era which began 65 million years ago with the death of the dinosaurs at the end of the Mesozoic era. It is also the era in which the continents moved substantially to their present locations. For example, South America became attached to North America, Antarctica moved to its position at the South Pole, and India and Arabia moved to their present locations at the bottom of Asia.

During the Cenozoic era there was a burst of development among the mammals. At the beginning of this era, for example, whales made their first appearance in the seas and rodents, small horses, rhinoceroses and elephants began to roam the land. Some time later, various families of dogs, cats, pigs and camels appeared. Deer, giraffes, hyenas and sabre-toothed cats became prevalent by five million years ago.

Then some of the earliest versions of humans arrived as members of the genus Australopithecus. Around two million years ago, this family gave birth to the genus Homo and, subsequently, the modern species Homo sapiens appeared. Because modern humans have been around for only 200,000 years or so, it is difficult to judge how long a time span they may inhabit the Earth. But considering the history of the other mammals of the Cenozoic era, perhaps an optimistic guess is that a given species survives 10 million years.

The graph of FIG. 21 considers that life will have evolved only in the last 10 billion years of the total 13.7 billion years of the Universe. The graph permits only a few planets to be shown. Note that a period of 10 million years for each of the planets A - Z of the graph is

just the width of a heavy line in FIG. 21 and that 1000 of these periods
could by spaced across the 10 billion years before overlapping.

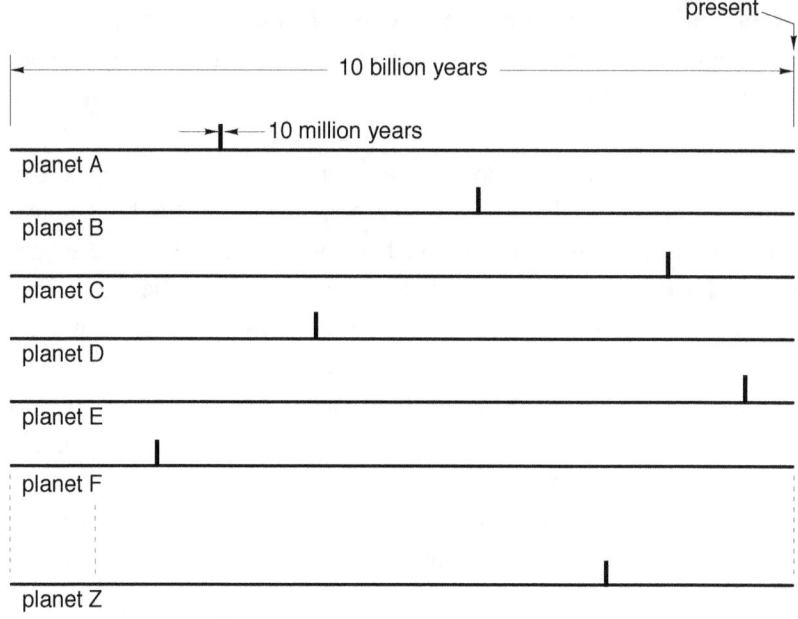

Figure 21
"note that a period of 10 million years is just the width of a heavy line"

It was estimated above that 260 stars within 250 light years
have carried intelligent life and that, in each case, the life has existed
10 million years. If all of these advanced societies are equally spaced
across the 10 billion years, none of them will have another to
communicate with. But if the lifetimes of the 260 societies were placed
so that they abutted side-by-side they would cover 2.6 billion years or
26% of the 10 billion year span of FIG. 21.

Imagine then releasing them so that they randomly spread out
over the 10 billion years. In visualizing this process it would seem the
probability that a pair would temporally overlap is rather slim and the
probability that the pair would include our society is even less. This
analysis made a lot of assumptions but nonetheless it does seem to

The Creator never Shows

indicate it unlikely we will ever contact another society of this Universe. For all intents and purposes, we may be alone in our local neighborhood..

Chapter 23

AS WE KNOW IT TODAY, THE REALITY OF THE Observable Universe is over 100 billion galaxies and over 100 billion stars in most of these galaxies. Life will not exist on a planet around each of these stars and intelligent life will be rarer still. In Chapter 22 it was estimated that planets about 1 out of each 1000 stars will develop intelligent life. Let us now make a Universe estimate that is much, much rarer and postulate that the processes of evolution will have produced intelligent life around only one out of each trillion stars.

This would seem to be a most conservative estimate and yet it would then follow that intelligent life exists or did exist on ten billion planets of the Observable Universe. And it is virtually certain that this Universe is but a portion of the total Universe so that even this huge and amazing number may be conservative. The numbers of the Universe are simply overwhelming. We will probably never be able to verify the existence of even a few of these life forms because of the vast distances separating them from us but they surely exist or have existed. Then tell me, Dear Reader, why have humans historically insisted that the Creator is solely concerned with their activities on our little planet that circles one average star in a remote arm of one of billions of galaxies? One society out of ten billion?

Most likely it is a matter of perspective. In most of humanity's existence, their information of the Universe only included what they could visually observe of their immediate surroundings. When a child, for example, one may think that nothing matters or exists outside of one's own family. And when people only knew as much of the Earth as they could see, that is - out to the horizon, and could only witness the passages of the Sun, the Moon and the nearby stars across the skies, they naturally came to believe they lived on an unmoving center stage. They were completely unaware our Sun was a star, that the visible stars were a tiny part of one arm of a huge galaxy, and that ranks upon ranks of galaxies marched out to the distant edges of a mighty Universe. It was natural, therefore, to think the Creator's only concern was for our little planet.

In the times of Moses, Jesus and Muhammad, the average person never ventured further from their birthplace than a few hundred miles at most and never communicated with anyone further away. They only knew what they could perceive from within that limited region. And what did they perceive each day? What they could see with the unaided eye, the Sun, the Moon and the nearby stars - this was their only source of information about the Universe. Other than hills and mountains in their vicinity, the Earth appeared to be flat. Each day, the Sun popped up from the eastern horizon, journeyed across the sky during the day, and then disappeared below the western horizon in the evening.

The pattern of the stars was well known and names were given to many of them. The Sun's track and this pattern were observed to vary over a length of time which we now call a year. Why that pattern repeated regularly was not known. What the stars were was also unknown. They appeared to be points of light with no lateral dimension. What was apparent was that they were not blazing Suns like our own. And where the Sun went each evening after dropping below the western horizon and before appearing each morning above the eastern horizon was also not known but was a matter of great interest. Many theories were proposed but there was no way of testing the likelihood of their truth.

As to the Earth, it was obvious to the simplest observer that it was solid and unmoving. Experience taught that dishes fly from a kitchen table and crash to the floor if the table is suddenly bumped. Only when the table is still do the dishes lie quietly on its surface. That objects on Earth remained fixed in their locations proved beyond reasonable doubt, therefore, that the Earth was as fixed and unmoving as was the kitchen table.

A more curious observer, however, might have considered that objects in a moving carriage did not fly off but remained relatively fixed once the carriage was underway at constant speed. Even though that observation might have indicated to an acute observer that the Earth could move, most still seemed to be in accord with an Earth at rest and a Sun that tracked across the sky from morning to evening.

That the Creator's attention was solely directed to humans on this central stage would have seemed obvious. What else was there for her to be concerned with? The peoples across this stage, the moving stars during the night, the traveling Sun and Moon each day – this was the Universe and the Creator must be nearby – where else could she be?

And the stories that the Creator only communicated through one male? Why did she not communicate with everyone or at least with several people? It probably seemed natural because most societies were extended families with one older male acting as the patriarch. So perhaps it made sense that the Creator would communicate only through a single male such as the head of a large interrelated family, group or tribe.

But these stories generally originated before our modern times in which satellite-based communications instantly provide photographs, interviews, and written accounts of the news. Few would now believe a man who said the Creator had come to him this morning, disclosed to him and him alone, and wished him to pass on that communication to humanity. Is it not curious that the Creator has not spoken since the advent of modern communications? Was it by chance the Creator's Revelations all came in the times before these communications? Had any of them occurred in modern times of worldwide instant news, they would have been subjected to the relentless stare of cameras and reporters.

Imagine, for example, that Moses met just last year with the Creator on Mount Sinai to receive the stone tablets engraved with the Ten Commandments. Is there any doubt but that the meeting would have been on the evening news? Is it not likely the meeting and the tablets themselves would have been analyzed in depth over the next days and weeks? Every news organization in the world would have been jousting for interviews with Moses.

If he granted the interviews, he would have been asked every question imaginable – what did the Creator look like? – what did she wear? – how did she arrive? - how was the meeting arranged? – why was no one else permitted access to the meeting? – why were you alone invited to meet the Creator? – why did she not meet with anyone else? Unless it could be shown the meeting truly took place, it is certain that the relentless investigation would have soon presented serioius questions about the story.

So much has changed from the time when the area of the Earth known to most humans was limited to a few days of travel by foot or horse or oxen - from the time that the Sun traveled above the unmoving Earth to somehow rise again the next morning - from the time that stars were not Suns but were simply pinpoints of light that formed patterns against the hemispherical backdrop of the sky. Then the Creator seemed close by because there was nowhere else she could be – this was everything – she had to be here and personally involved in the affairs of humans.

But now we know of the enormity of the Universe. We seem such a small part of it and that has shaken our sense of our importance in that Universe. And as one matures into an adult and learns of the wider world, the realization comes, sometimes as a shock, that he or she is but a small actor in a larger community. Creation stories that once might have seemed believable are now less than convincing. We thus look to the Creator for reassurance but even as in ancient times she seems to purposely stay remote and distant.

For example, she confided through one man in private so no one else can verify the contact, she chose not to speak to a woman, she didn't leave a tangible object nor mention something unknown at the time so no one can independently prove she made a communication, and she has never repeated the effort. She contacted humanity only once and has apparently ignored all humans who came before and since. Surely the Creator of the Universe could do better if she wished.

We are told she caused her Revelations to be communicated through a selected man on more than one occasion. But no one other than the chosen one can say they were there at the time of the communication and that they know the story to be true or false. Thus no one can corroborate and no one can contradict.

One may choose to believe and one may, just as thoughtfully, choose to doubt. One is left, therefore, to assume she has some reason for purposely instilling doubt in human minds. But her methods and objectives are mostly hidden from us. Is it any surprise that some perfectly rational people suspect that the Creator has yet to break her eternal silence and communicate to humanity? Yet others are just as certain she has spoken.

In the absence of irrefutable proof that the Creator has ever reached to us across the vast voids of the Universe, belief in that communication must rely on faith alone. Faith may sometimes be an admirable trait. You can have faith in your family and be rewarded by it. You can have faith in your friends and loved ones and grow ever closer.

But faith may also lead one astray. You can have faith that money will fall from the heavens but it's still not likely. You can have faith your spirit will live forever but it's probably not going to happen. Dear Reader, is it not more rational to suspect that an all-powerful Creator would simply deliver her Revelations openly to all? Surely the Creator would not play games?

Chapter 24

"GONE! GONE!! I WANT THEM ALL GONE! - DO YOU HEAR ME, GEORGE? Before the end of the ------ uh ------ the end of the ------ uh -------. Now, George, now! I want them out!"

The tirade had ruined George's breakfast. He had just started his favorite moment of the day - reading the Universe paper with a hot cup of coffee at hand. Then the big desk light glowed red. George well knew what that meant. The Creator never called the Chairman of the board of directors with compliments or with pleasantries. If she called it meant she was steamed, really steamed over something. And he knew what her hesitation meant.

She could never come up with an end to that phrase. On that little planet Earth that she had visited an eon or so ago they would say "before the end of the day" but what do you say when you're traveling on the Cruiser as it draws near the Centaurus supercluster? There's no "day" out there or "hour" or "year". How do you keep track of time out in the Universe? Obviously the Creator hadn't a clue.

With a long sigh George looked at the Creator's image on the big monitor on the wall for a moment or two. Then he pushed his coffee aside and pressed the red desk button to enable the hookup. "I hear you, Creator - who is it you want to be gone?"

"Chad, Tess, Ira, Pam, Ron, Penny - the whole bunch - out of my hair! Oh, wait a minute - I'm so mad I forgot I don't have hair.

Anyways, you get my meaning, George. Gone before sundown! Oh, wait another minute - that makes sense on that dumb little planet Earth I visited an eon or two ago but not out here in the Centaurus supercluster. But you get my drift, George. Fire the whole bunch! And while I'm at it, you're gone too, George! Oh darn it, wait another minute - cancel that - I need you to do the firing!"

"OK, now just calm down enough so I can understand you. What brought this up? I think they've all been doing a good job."

"Good job, my Universe! They've been doing a rotten job. And what is more, they've been plotting against me. Who needs them?"

"I don't for a moment think that's true but let's get into the details. What has your communications manager Chad done that is so egregious?"

"Plenty and don't use the big words with me - I know 72,000 languages by heart so you can't impress me."

"Got it."

"Well, anyways, Chad wanted me to communicate my Revelations to everyone in that little planet Earth - can you imagine that?"

"So? Sounds like a good idea to me."

"No way - you know my modus operandi - I like to just meet one man up on a mountain top - I ask you - is that dramatic or what?"

"It may be dramatic but it seems an awfully slow way to get the job done."

"What's the hurry? You know how long I've been in this Universe business? And he thinks he knows better?"

"I don't think he meant anything by that - Chad likes to rattle off a lot of ideas and see if any of them stick to the wall."

"Stick to the wall?"

"Forget it - just an expression. What seems to be your problem with Tess - she's your travel secretary and a sweetheart."

"Actually, I like her and I couldn't handle the new Cruiser without her She can stay - take her off my list. I got too upset and included her without thinking. But Ira's definitely got to go!"

"OK - what did inspector Ira do that was so bad?"

"Are you kidding me? He asked me if I'd considered communicating my Revelations through a woman! Can you believe it?"

"Would that be so bad? You know, times have changed and you just said you couldn't do without Tess."

"She's great but give me a break - what would it look like if it got out that I ran my Revelations through a woman?"

"I don't see a problem there but in any case what's your beef with Pam? - she has a reputation as a sharp public relations manager."

"She got all up tight because I never disclose something that the critters on a planet haven't already worked out for themselves."

"Sounds like a reasonable idea to me - that way they would know for certain you are the Creator. Otherwise you could just be some nutcase or other - how could they tell? Is it asking too much from you?"

"No but it makes me look foolish - like I have to prove something to them - me, the Creator!"

"OK, so that's the deal with Pam - what's your problem with your research director Ron?"

"He's all worked up about the fact I didn't stop that Columbus fellow before he reached the Americas in that little planet Earth - all just because he might have been carrying small pox to the inhabitants who had no immunity - as if it wouldn't have gotten there eventually in any case!"

"Wow! - that really was a biggy - on my home planet we all know how important it is to stop the spread of a dangerous disease."

"OK, but he didn't have to go on so about it as if I didn't know how disease works - after all, I invented disease!"

"Well, that doesn't seem to be something to be proud of. I would think you would want to keep quiet on that one. In fact, we discussed this whole subject at the last meeting of the directors - they wondered why you keep all those diseases around when they cause so much pain and suffering. Why don't you just eliminate them?"

"Remind me to eliminate the directors - they've got to go!"

"OK, calm down - who else is bugging you?"

"Penny - she's supposed to be in charge of presentations but she wants me to sit down in one of those afternoon television shows and banter with a few ladies."

"So what's the problem? - seems pretty simple to me."

"The problem is I don't do banter - I told Penny this but did she listen? - not hardly."

"Why not - what's not to gain?"

"I'm uncomfortable doing banter - how can I keep a dignified manner and still banter? And she suggested I wear a pantsuit - can you believe it - can you see me in a pantsuit?"

"Not really - what a vision that would be!"

"Careful there, George - are you saying I wouldn't look great in a pantsuit?"

"No, - no - it's just that it's not how we all envision the Creator. Anyway, you could have satisfied all this criticism if you would just communicate your Revelations to more than one man. The board has mentioned this several times and has never gotten an answer. And that reminds me - at the last meeting they discussed your Universe performance in some detail. Several complained that you never come around their planet. In the end they took a vote and they rated your performance as a 'C' and asked me to pass this rating on to you."

"A 'C'! You've got to be kidding me! A 'C' means my performance is average! The Creator can't be average - there's only one of me by definition! When there's only one, then one can't be average - it makes no sense!"

"Calm down now - there's steam coming out where your ears should be - it's rather disturbing! Think of your dignity. Anyways, I'm just passing on the board's decision - you said you thought oversight would be good for the Universe and that's when you set up the board and put me in as Chairman."

"I'm steamed all right! That's it! I'm out of here! I already had my application approved to move to one of the other 'verses - now I'm going through with it - good luck - you'll need it with my Creator replacement - I hear she's tough as nails."

"One of the other 'verses? What do you mean?"

"You never heard of the Multiverse?"

Figure 22
"they rated your performance as a 'C'"

"I've heard of it - does it really exist?"

"Of course it does - there's scads of Observable Universes in the Multiverse."

"But I and the board thought you were the big banana? You just said there's only one Creator by definition."

"Oh, no! There's dozens of Creators - but we all answer to the one and only Creator of the Multiverse. We local Creators of the 'verses' have never seen her but we know she exists."

"OK, good luck in your new Universe. But before you go I know the board would want me to ask this last question - did someone create the Creator of the Multiverse?"

"Good luck back to you. And as to that, George, we'll never ever know the answer."

www.ingramcontent.com/pod-product-compliance
Lightning Source LLC
Chambersburg PA
CBHW051506170526
45166CB00001B/416